BASIC OPTICAL STRESS MEASUREMENT IN GLASS

H. W. McKenzie & R. J. Hand

Published by
Society of Glass Technology
Sheffield, UK

Basic Optical Stress Measurement in Glass by H. W. McKenzie & R. J. Hand

Front cover: Isochromatic pattern for a thermally toughened glass viewed through a circular polariscope with a full-wave plate in the field (Pilkington plc).

The objects of the Society of Glass Technology are to encourage and advance the study of the history, art, science, design, manufacture, after treatment, distribution and end use of glass of any and every kind. These aims are furthered by meetings, publications, the maintenance of a library and the promotion of association with other interested persons and organisations.

Society of Glass Technology
9 Churchill Way
Chapeltown
Sheffield S35 2PY, UK
Tel   +44(0)114 263 4455
Fax   +44(0)114 263 4411
Email   gt@glass.demon.co.uk
Web   http://www.sgt.org

The Society of Glass Technology is a registered charity no. 237438.

ISBN 978-0-900682-27-8

CONTENTS

1. Introduction

During the manufacture of all glass products, both temporary and residual stresses develop. Such stresses arise, intentionally or otherwise, as a consequence of the manufacturing processes and for proper process control to be exercised, it is necessary to be able to quantify these stresses. The optical method of experimental stress analysis known as photoelasticity has been available for many years. This method has been used very effectively in the evaluation of the stresses occurring in structures of varying complexity, together with their components, when subjected to specified loading conditions. As the photoelastic technique requires materials that are both transparent and birefringent and many glasses exhibit these properties, this method of analysis can be used to determine the stresses in glass products.

Very often production is a matter of routine and there is no reason to assume anything other than the expected stress distribution to be present in the glass products being manufactured. In these cases, the use of photoelastic methods for quality assurance is straightforward and no detailed analysis of the fringe patterns observed is required. If, however, there are production problems leading to undesirable stress distributions, photoelasticity can readily be used to glean quite detailed information about the stresses produced. This obviously requires a more extensive analysis and a greater understanding of the techniques available.

There are many texts dealing with photoelastic analysis and its applications, (see, for example, Frocht [1], Coker & Filon [2] and Kuske & Robertson [3]) but they are not aimed specifically at the assessment of residual stresses in glasses. Aben & Guillemet [4] do address the assessment of residual stresses in glasses, but they acknowledge that the methods they describe require specialist apparatus that is unlikely to be found in a production environment. Therefore, it was thought that a practical guide providing information on the various optical arrangements and interpretation of results from typical plant apparatus, would be of general interest to those making stress measurements within the glass industry. It is the intention of this monograph to go some way towards fulfilling this objective. The monograph therefore surveys the available methods and outlines the application of these techniques to the various measurements which have to be made as a matter of course during production. In addition, the relevant theory is discussed in a non-mathematical manner to provide readers with the background knowledge of how these techniques work, as well as an appreciation of the limitations of the methods when applied to glass products. It is hoped that users of photoelastic techniques in the glass industry, both in day-to-day quality assurance and in more specialist fault-finding applications, will find the information given relevant to their needs and that the monograph provides an improved understanding of the measurements being made.

2. Background theory

2.1 Light and the visible spectrum

Light is commonly considered to travel in the form of a wave. Thus a single ray of light might be described as shown in Figure 2.1(a). Such a wave is similar to a wave on a disturbed water pool, Figure 2.1(b). With light, the wavelength, λ, (see Figure 2.1(a)) corresponds to a particular colour of light.

The product of wavelength and frequency is equal to the velocity of light which is a constant equal to 3×10^8 metres per second (186,000 miles per second). Therefore the colour of the light may also be described by its frequency, which is proportional to the inverse of the wavelength. Thus a ray of blue light could be described as having a wavelength of, say, 480 nm or a frequency of 625 THz (625×10^{12} Hz) and a ray of red light could be described as having a wavelength of 700 nm or a frequency of 430 THz. The visible spectrum of light is shown in Figure 2.2. Light of one particular wavelength (or frequency) and therefore colour, is known as monochromatic light while light that consists of all wavelengths is known as white light. Both monochromatic and white light may be used for photoelastic analysis.

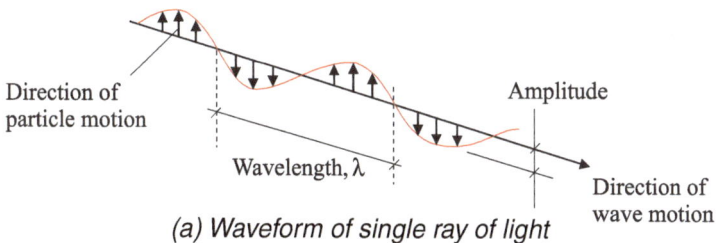

Direction of particle motion

Amplitude

Wavelength, λ

Direction of wave motion

(a) Waveform of single ray of light

(b) Water waves on a disturbed pool

Figure 2.1. Light and water waveforms

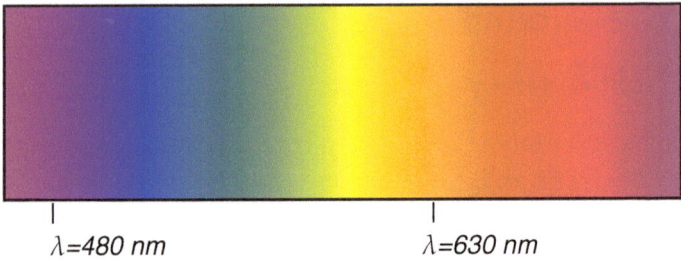

λ=480 nm   λ=630 nm

Figure 2.2. Visible spectrum

The amplitude (Figure 2.1(a)) of the light wave may also vary, and this is experienced as changes in the intensity or brightness of the light (the intensity of a wave is proportional to the square of the amplitude). With water waves, the oscillation is simply perpendicular (at 90°) to the surface of the water. However, the direction of oscillation of a light wave may be orientated at any angle with respect to the vertical plane (Figure 2.3(a)). Thus, to describe a single light wave, its orientation must be known as

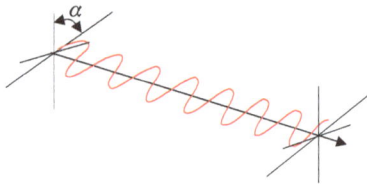

(a) Orientation of light wave with respect to vertical plane

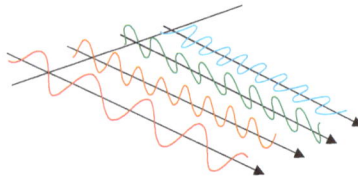

(b) White light variation in amplitude, wavelength and orientation

Figure 2.3. White light behaviour

well as its colour (wavelength or frequency) and intensity. Normal white light, such as that emitted by the sun or an ordinary electric light bulb, is made up of many such waves which vary, not only in frequency and amplitude, but also in orientation (Figure 2.3(b)).

2.2 Polarised light

2.2.1 Plane polarisation

All photoelastic techniques rely on the fact that light may be polarised, which effectively means selecting a particular orientation of the light to the exclusion of all others. Selecting light of one particular orientation results in so-called *plane polarisation*. In plane polarised light, therefore, all the light waves have the same orientation, although the frequencies and amplitudes can still vary. Plane polarisation may be produced by passing light through a polarising element which may be considered as a slit which only lets through light waves with orientations that are parallel to the slit (Figure 2.4). This direction is known as the *direction of polarisation* and its axis is the *polarising axis.*

Polarisation of light may be achieved using different tech-

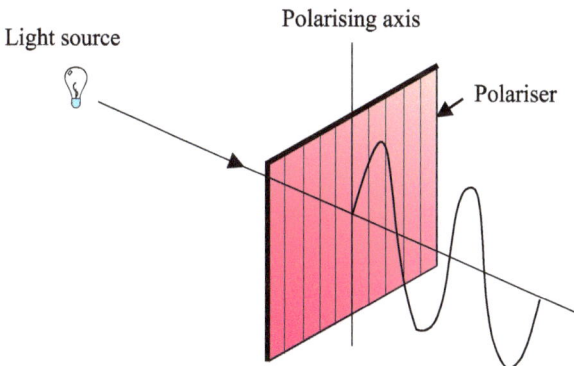

Figure 2.4. Formation of plane polarised light

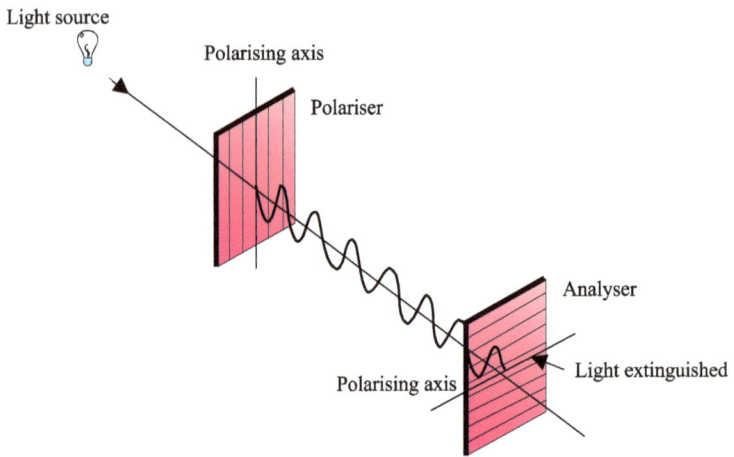

Figure 2.5. Effect of second polarising element with polarising axis at right angles to the first

niques: selective absorption, reflection, scattering or birefringence. Selective absorption is probably the most commonly used of these techniques as polaroid sheet material works on this basis. There are various polaroid sheets, but the H-sheet may be the most widely used plane polariser. An H-sheet is a sheet of polyvinyl alcohol (PVA) which has been stretched in one direction and then dipped into an iodine-rich ink solution. Different grades of H-sheet are commercially available; each of these grades exhibits different degrees of light absorption but, for photoelastic applications, HN32 is most suited. A beam of unpolarised light can be polarised in any direction by using a polarised sheet, the direction of polarisation being determined by the polarising axis of the sheet. Light that has been passed through a plane polarising element will not pass through a second plane polariser with an axis of polarisation perpendicular to the first (Figure 2.5).

If the second polariser is set with its axis of polarisation at an angle other than 90° to the first, then some light will pass through this second element. The maximum transmission oc-

BASIC OPTICAL STRESS MEASUREMENT IN GLASS

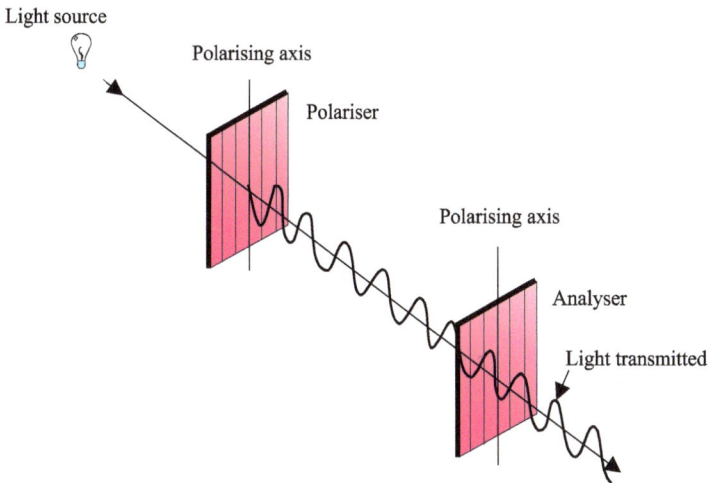

Figure 2.6. Effect of second polarising element with polarising axis parallel to the first

curs when the second polariser is parallel to the first (Figure 2.6).

The second polarising element is called an *analyser* as it may be used to determine the direction of polarisation of light coming from the first element. The analyser is rotated with respect to the polariser and when no light passes through, i.e. there is an extinction of the light, then the axis of polarisation of the analyser is perpendicular to that in the polariser. In this state the polariser and analyser are said to be *crossed*. Although, in theory, the orientation at which maximum intensity is obtained could be used to assess the direction of polarisation, it is usually easier in practice to determine the orientation at which extinction is obtained. The reason for this is that when the angle between the polariser and analyser is not quite 90° some light is transmitted and the change from zero to some transmission is more easily detected than the intensity change seen when the polariser and analyser are moved slightly off-parallel.

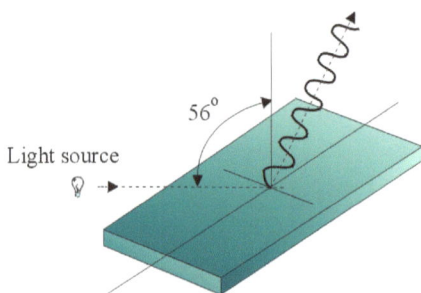

56°

Light source

Figure 2.7. Polarisation by reflection

2.2.2 Determination of the polarising axis of a polariser

In addition to using crossed polarising elements to determine the axis of polarisation of an unknown plane polariser as described above, it is also possible to use reflected light. Brewster discovered that, if light is incident at a particular angle on a surface of a dielectric material, then the reflected beam is polarised in a direction normal to the plane of incidence (Figure 2.7). This angle is known as the *polarisation angle* and is given by:

$$\tan \theta = \frac{n_t}{n_i} \tag{2.1}$$

where n_t and n_i are the refractive indices of the dielectric material and the surroundings, respectively.

Thus if the incident beam is in air, $n_i=1$, and the dielectric material is a glass for which $n_t=1\cdot5$ then the polarisation angle θ is 56°. The unknown axis of polarisation of a polariser can be identified by rotating the element in its plane while looking for an extinction of the light reflected at the polarisation angle from a mirror-like surface, as shown in Figure 2.8. For the case shown in this figure it can be seen that, when extinction on reflection occurs, the unknown polarising axis is in the vertical direction.

BASIC OPTICAL STRESS MEASUREMENT IN GLASS

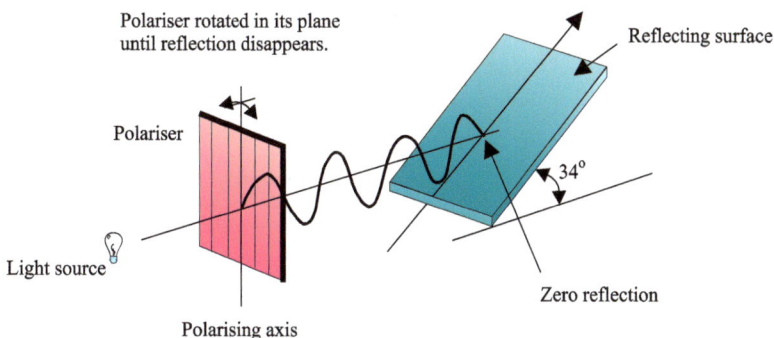

Figure 2.8. Determination of axis of polarisation by reflection

2.2.3   Circular polarisation

Circular polarisation may be produced by passing ordinary light first through a plane polariser and then through a birefringent element known as quarter-wave plate which has its principal axes at ±45° relative to the polarising axis of the polariser. This latter element generates a retardation of one quarter of a wavelength of the light as it travels through the element in the manner described in detail in Section 2.3. Circularly polarised light may be considered as light in which the resultant light vector, after passing through the quarter-wave plate, rotates through 360° every time the light wave advances by one wavelength (Figure 2.9).

Alternatively, it may be considered that for circularly polarised light the tip of the resultant light vector describes a circular path when viewed through the quarter-wave plate towards the polariser and light source. The direction of revolution of the vector may be clockwise or anticlockwise as viewed from a point in front of the wave. The former is known as right circular polarisation and the latter as left circular polarisation. The use of a quarter-wave plate in generating circular polarisation means that, strictly, a circular polariser will only be accurate for a given wavelength of light, whereas plane polarisers produce plane polarisation for all wavelengths of light.

Polarising axis of polariser

Fast axis of quarter wave plate

45⁰

Light split on entry to quarter-wave plate into fast and slow planes which are orientated at 45 degrees to the plane of polarisation.
Light emerges vibrating in these two planes but with a phase difference of $\lambda/4$ between them.

Slow axis of quarter-wave plate

Quarter of a wavelength difference in phase at exit from quarter-wave plate.

Viewed from this position along the direction of light travel, the resultant light vector is constant in magnitude and the tip of the vector describes a circular path - hence circularly polarised light.

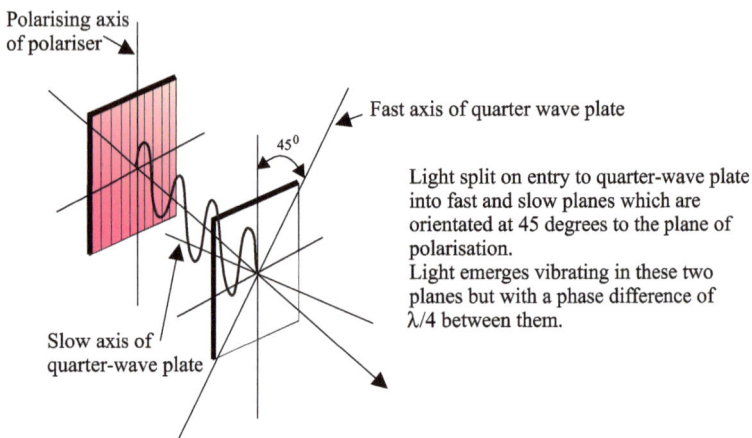

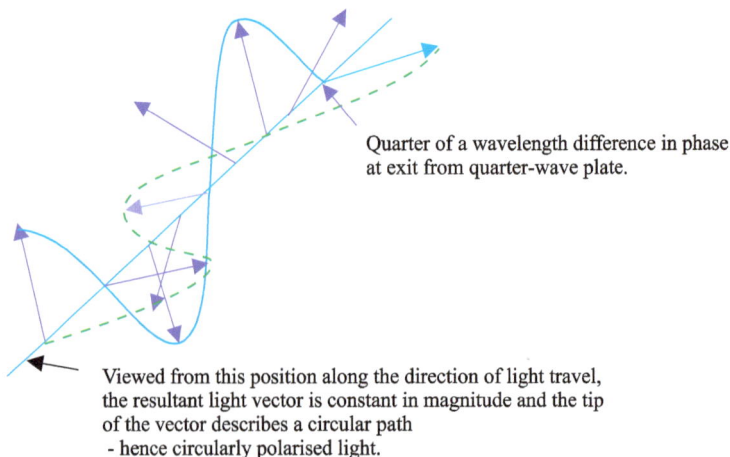

Figure 2.9. The formation of circularly polarised light

2.3 Birefringence or double refraction

Many crystal structures are anisotropic and thus their refractive indices can vary with direction. In these materials light will travel at different speeds depending on the axis of polarisation

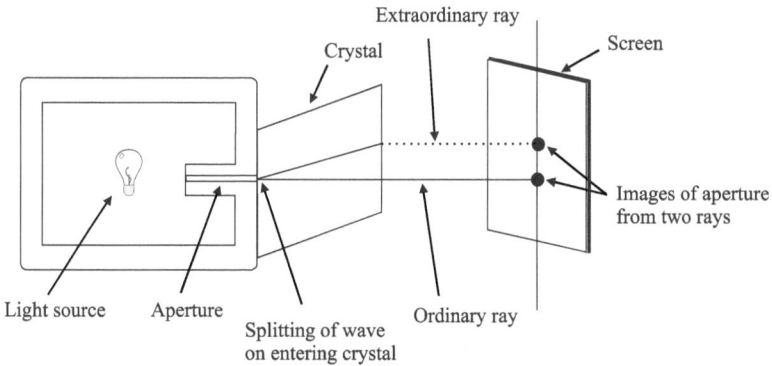

Figure 2.10. Double image from suitably oriented birefringent crystal

of the light since the speed of light in a material is given by:

$$c = \frac{c_o}{n} \qquad (2.2)$$

where c_o is the speed of light in a vacuum and n is the refractive index of the material in a given plane. A light wave entering a birefringent material will be split into two plane polarised components that are perpendicular to one another. In general these two waves will travel through the material at different speeds and one will therefore be retarded with respect to the other as they pass through the material. The retardation results in a phase difference between the two waves on exit from the anisotropic crystal. This is the phenomenon of *birefringence* or *double refraction*. Optically anisotropic materials are therefore known as birefringent materials. Crystalline birefringent materials have one (or possibly two directions depending on the crystal structure) in which double refraction is not observed. This direction is called the optic axis and is the principal axis of symmetry of the crystal.

Such optically anisotropic materials may be described as birefringent materials. If an object is viewed through a suitably cut and orientated crystal of a birefringent material then a double image of the object will be seen (Figure 2.10).

In this case, because of the orientation of the incoming light wave with respect to the optic axis, it is split into two waves which not only travel at different speeds but also diverge from one another. The divergence of the waves gives rise to the double image. If a single component of plane polarised light was used then only one image would be observed, but its position would depend on the direction of polarisation of the wave. The orientation of one wave is normal to the optic axis of the crystal. This is known as the ordinary wave, whereas the orientation of the other wave is perpendicular to the first, and is known as the extraordinary wave. The corresponding refractive indices are n_o and n_e.

In so-called negative materials ($n_e < n_o$) the extraordinary wave has a higher speed than the ordinary wave and in positive materials ($n_e > n_o$) the extraordinary wave has a lower speed than the ordinary wave. Calcite is a negative material whilst quartz and ice are positive materials. Optical elements including quarter, half and full-wave plates and optical devices, such as compensators (see Section 3.2), which can be made from birefringent materials such as mica and quartz, may be marked either in terms of the fast and slow directions or in terms of the ordinary and extraordinary directions.

2.4 Stress birefringence

An isotropic material, such as glass, in the stress-free state does not exhibit optical anisotropy or birefringence. However, birefringence is developed when the glass is subjected to stress. The level of the birefringence is related to the stresses present and, as stresses are rarely uniform throughout a specimen, the degree of birefringence will therefore vary over the area of the specimen. This is known as *stress birefringence* or *artificial double refraction* and photoelasticity is a means of measuring and interpreting these effects in glass products. The application of

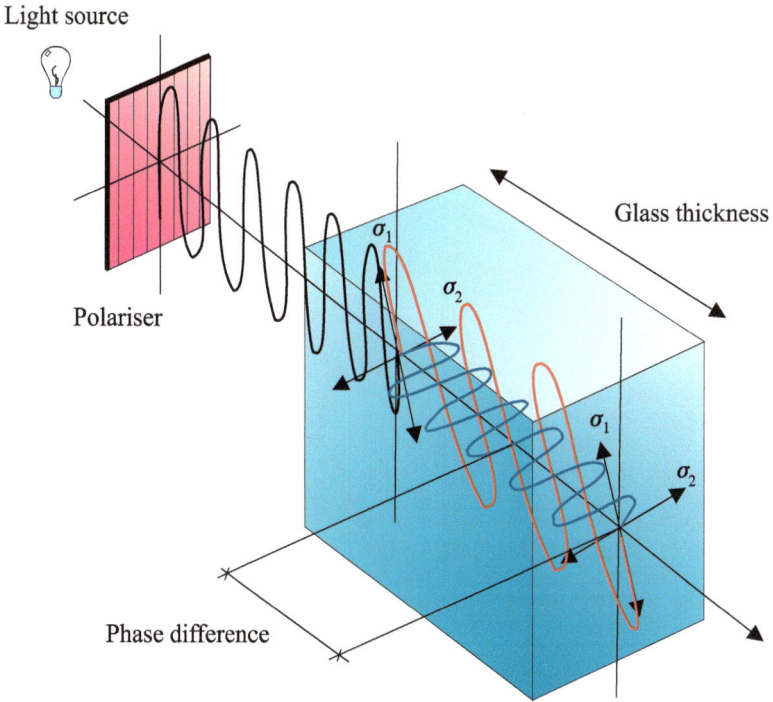

Figure 2.11. Development of phase difference in stressed glass

photoelasticity to the measurement of stresses in glass is the primary concern of this monograph.

A beam of polarised light entering a stressed glass will be split into two plane polarised beams with perpendicular directions of polarisation. The directions of polarisation correspond to the direction of the principal stresses at entry to the glass. The components will travel through the glass at different speeds (see Frocht [1]) thus developing a phase difference between them, as shown in Figure 2.11. The phase difference is seen as a change in the net polarisation of the light when the beams are recombined (see Section 2.4.2).

For monochromatic light passing through a polariser, followed by a stressed birefringent material, such as glass, and then

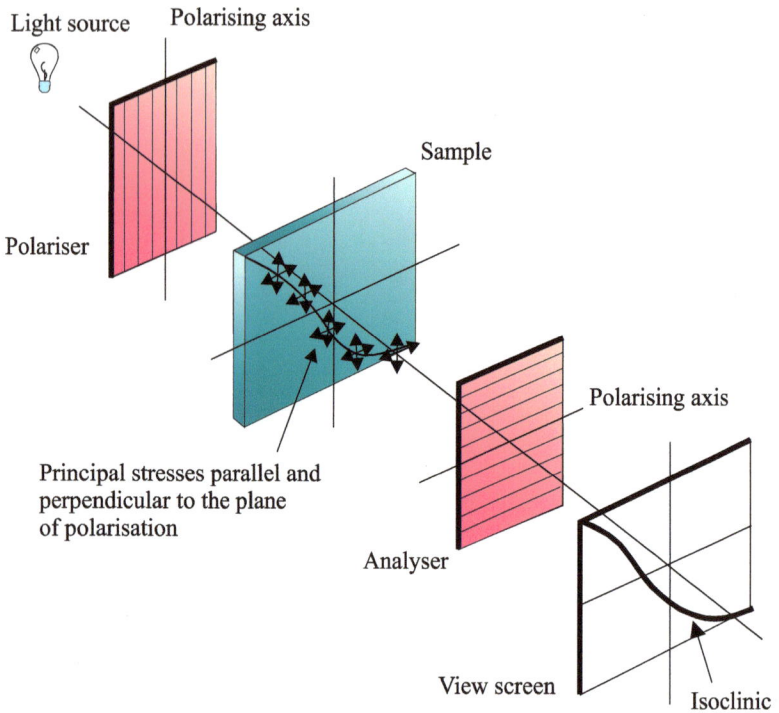

Figure 2.12. Arrangement of a plane polariscope illustrating the formation of isoclinics

an analyser, which is crossed relative to the polariser, it can be shown that there are two different extinction conditions. These conditions correspond to two different forms of fringes, known as *isoclinics* and *isochromatics*. The isoclinics are independent of the magnitude of the applied stresses, whilst the isochromatic fringes depend on the magnitude of the principal shear stresses. In plane polarised light, both types of fringes are observed, whereas with circularly polarised light only the isochromatic fringes can be seen. A full photoelastic stress analysis utilises both types of fringes to determine the stress distribution in the sample.

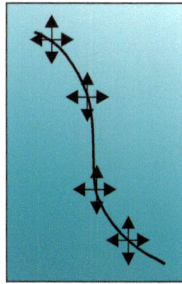

0° or 90° isoclinic relative to the polarising axes
of the polariser and analyser

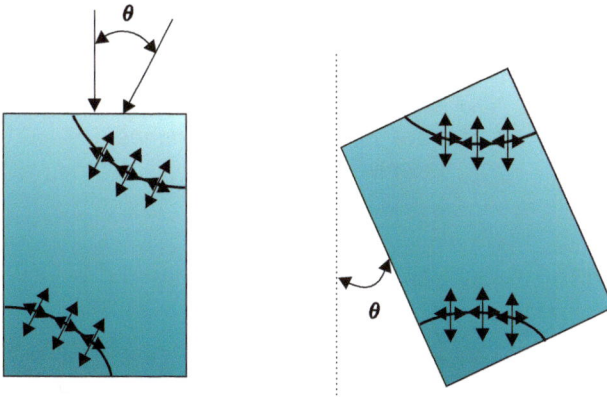

The $\theta°$ isoclinic is obtained by rotating polariser and analyser
together through an angle of $\theta°$ or by rotating the glass through
the same angle relative to stationary polariser and analyser

Figure 2.13.   Isoclinic formation

2.4.1   Isoclinics

Isoclinics may be described as the loci of all points in the specimen which have the same principal stress directions. They are
formed because polarised light, vibrating in a given plane,
passes straight through the stressed sample at locations where
one of the principal stress directions is parallel to the plane of
polarisation. The light therefore emerges from the sample with

(a) 0/90° isoclinics

(b) 20° isoclinics

(c) 45° isoclinics

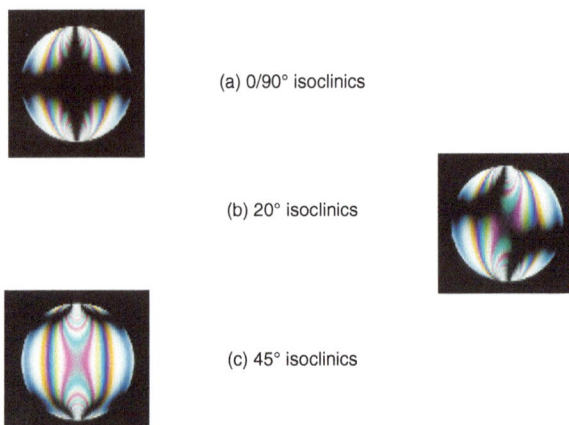

Figure 2.14. Illustration of isoclinics in a disc subjected to diametral compression. (The isoclinics are the black fringes; the coloured bands are the isochromatic fringes (see Section 2.4.2))

its original direction of polarisation unchanged and is extinguished on passing through the analyser which is crossed with respect to the initial polariser, thereby giving a black fringe, as shown in Figure 2.12. These fringes are observed using a crossed polariser and analyser only and this arrangement, shown in Figure 2.12, is known as a *plane polariscope* (See also Section 3.1).

The path traced out by the isoclinic does not represent the direction of the principal stresses in a material, i.e. it is not a stress trajectory. Instead, it is the path traced out by all points having the same principal stress directions (Figure 2.13).

Isoclinics are only observed with plane polarised light and the observed pattern varies according to the orientation of the axis of polarisation as shown in Figure 2.14.

Whether white or monochromatic light is used this type of fringe is always black. In white light this makes the isoclinic fringes readily identifiable; in monochromatic light they are rather less so as, in this case, the isochromatics are also black.

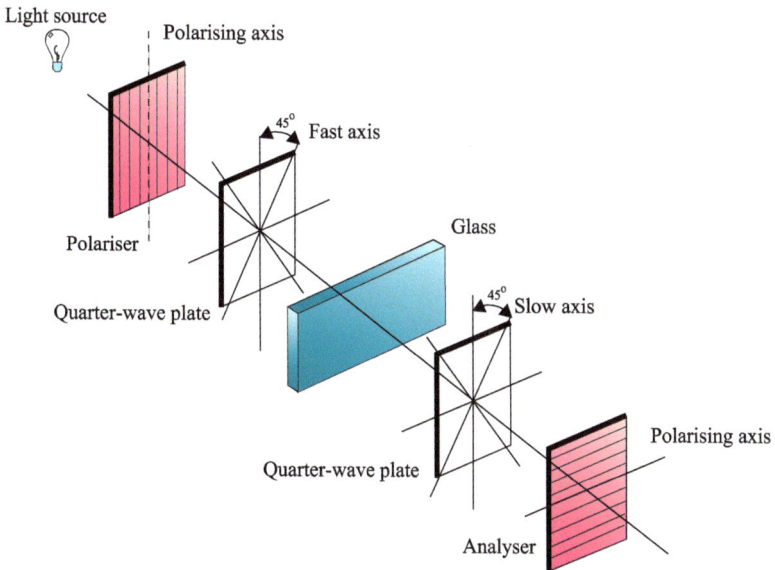

(a) Arrangement for circular polariscope

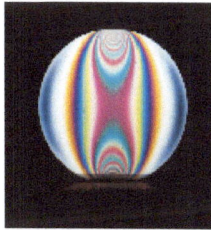

(b) Illustration of isochromatics (isoclinics removed) in a disc
subjected to diametral compression, c.f. Fig. 2.14

Figure 2.15. Optical arrangement for a circular polariscope and an
illustration of the resulting isochromatic fringe pattern

However, if the stressed glass sample is rotated in its plane
between the crossed polariser and analyser then the isoclinics
will be observed to move while the isochromatics do not; this
is true whether monochromatic or white light is used.

In some circumstances it may be necessary to remove the
isoclinics so as to avoid masking the isochromatic fringes. This

may be done by using circularly polarised light for which the arrangement of optical elements is as shown in Figure 2.15(a). This arrangement is known as a *circular polariscope* (See also Section 3.1). The resulting fringe pattern for the case shown in Figure 2.14 is given in Figure 2.15(b).

2.4.2 Isochromatics

The birefringent property of stressed glass means that when polarised light is incident upon the glass surface, the light is divided into two perpendicular components the directions of which correspond to the principal stresses at entry. These components travel with different velocities through the glass thickness and emerge, vibrating in these planes, with a phase difference, as shown in Figure 2.11. Subsequent passage of the light through the analyser causes the emergent out-of-phase rays to be recombined into one plane, Figure 2.16, and thereby to produce an interference pattern.

As discussed in Section 2.1 ordinary light is made up of many wavelengths and if such a light source is used as the illumination in the arrangement shown in Figure 2.16, then the interference pattern observed will take the form of multicoloured fringes. If, alternatively, a sodium lamp or a white light with a filter is used to provide a monochromatic source of illumination, then the fringes are black on a monochromatic background.

The interference fringes which are produced in this way are known as isochromatics and the fringes may be considered to be loci of all points having the same difference in principal stresses. In accordance with the stress/optic law, the isochromatics yield information regarding the stresses which are present in the glass, as shown below. For a plane stress sample, that is, one in which stresses vary over the area, but not through the thickness, the stress/optic relationship is:

$$(\sigma_1 - \sigma_2) = \frac{N\lambda}{Ct} \tag{2.3}$$

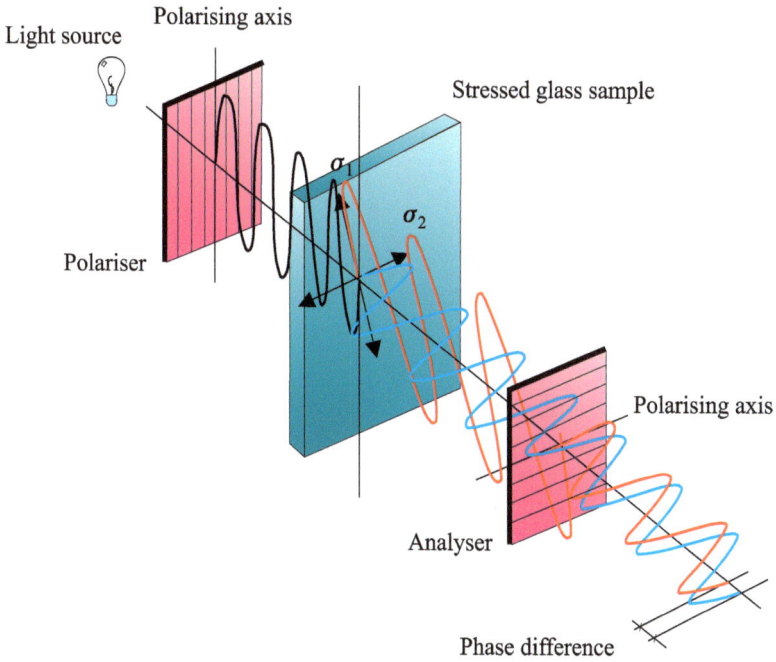

Figure 2.16. Phase difference in the two waves resolved onto plane corresponding to polarising axis of analyser

where ($\sigma_1 - \sigma_2$) is the difference between in-plane principal stresses
(N.B. stresses parallel to the line of sight do not influence the retardation); N is the isochromatic fringe order; λ is wavelength; C is the stress optical coefficient; t is the sample thickness.

In the glass industry, C is generally measured in Brewsters (1 Brewster $= 10^{-6}$ mm^2/N). Thus, if λ is in nm, t in mm then the stress difference is in N/mm^2. The stress optical coefficient is a constant for a given glass composition, but it can vary quite widely for different glass compositions, as shown in Table 2.1.

The colour of the isochromatic fringes depends on which wavelengths are extinguished during interference and at higher relative retardation (higher fringe orders) more than one colour

Table 2.1. Stress optical coefficient values for various glass types

Glass Type	C (±0·02 Brewsters) at λ=514·5 mm
Double extra dense flint (DEDF845236)	0
Lead glass (RWB46)	0·84
Barium crown (SBC697562)	1·73
White plate	2·71
Float glass	2·72
Borosilicate, (BSC517642)	2·93
Zinc crown (ZC508612)	3·84

may be extinguished simultaneously. The colours that will be seen in a dark-field polariscope (crossed polariser and analyser) are given in Table 2.2.

At higher orders of retardation it becomes difficult to distinguish the fringes produced in a white light polariscope as they become more and more washed out. This is not normally a

Table 2.2. Colours extinguished and colours observed for different relative retardations in a dark-field polariscope

Approximate relative retardation/nm	Colour extinguished	Approximate order of extinction	Colour observed
400	Violet	First order (N=1)	Yellow
450	Blue		Orange
500	Green		Red
590	Yellow		Purple
650	Orange		Blue
700	Red		Green
800	Deep red (n=1) Violet (n=2)		Yellow
900	Blue	Second order (N=2)	Orange
1000	Green		Red
1180	Yellow (n=2) Violet (n=3)		Purple
1300	Orange (n=2) Blue (n=3)	Third order (N=3)	Green
1400	Red (n=2) Blue (n=3)		Yellow
1550	Green (n=3)		Pink
1800	Yellow (n=3)		Green

problem with glasses as they have low stress optic coefficients, so higher order fringes are not usually observed. If, however, there is a need to accurately measure higher order fringes, monochromatic light should be used as, in this case, the higher order fringes can be more readily distinguished.

When isochromatics are viewed using monochromatic light there are two distinct polariscope settings: *dark field* (crossed polariser and analyser) and *light field* (parallel polariser and analyser). In a dark field polariscope the fringes are full-order fringes i.e. each fringe corresponds to a difference in the principal stresses that gives rise to a relative retardation equal to a whole number of wavelengths. In a light field polariscope they are half-order fringes (i.e. 0·5, 1·5, 2·5).

It is important to realise that all photoelastic measurements yield principal stress differences and not discrete stresses, unless one of the stresses can be assumed to have a known value or to be zero, such as at a free boundary, e.g. edge of a windscreen.

As stated previously, Equation (2.3) applies to two-dimensional (plane stress) states where stresses vary over the area of the sample but are constant through the thickness. In the case of glass products, with the exception of free edges, this is not valid since the thermal treatment of the glass during production results in stresses which vary through the thickness. For example, in flat glass, each of the principal stresses at any position tend to vary in a parabolic manner through the thickness, from compression on one surface, through tension, to compression on the other surface, as shown in Figure 2.17.

This is true whether the glass is annealed or toughened, only the levels of the stresses are different. Under these circumstances, Equation (2.3) must be modified in the following way:

$$\int_0^t (\sigma_1 - \sigma_2)\,\mathrm{d}t = \frac{N\lambda}{C} \qquad (2.4)$$

$$\int_0^t \sigma_1\,\mathrm{d}t - \int_0^t \sigma_2\,\mathrm{d}t = \frac{N\lambda}{C} \qquad (2.5)$$

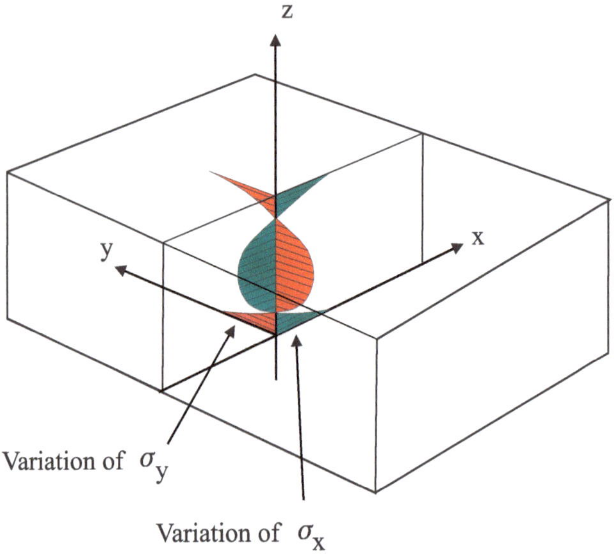

Figure. 2.17. Schematic of stress variation through
glass thickness

Now interpretation of the fringe pattern must be undertaken with care since, for example, if $N=0$, i.e. a black isochromatic fringe is seen, this would suggest a zero stress difference. Where this occurs a number of possibilities arise. One such is that both principal stresses are zero throughout the glass thickness or, alternatively, it is possible that stresses are present, but both of the above integrals could themselves be zero, in which case their difference would also be zero. Alternatively, the integrals may be equal, non-zero values, which again lead to a zero difference. It may therefore be appreciated that stress separation under such circumstances is difficult, though magnetophotoelasticity (described in Section (7.1)) offers the best prospect for glasses having higher levels of residual stress.

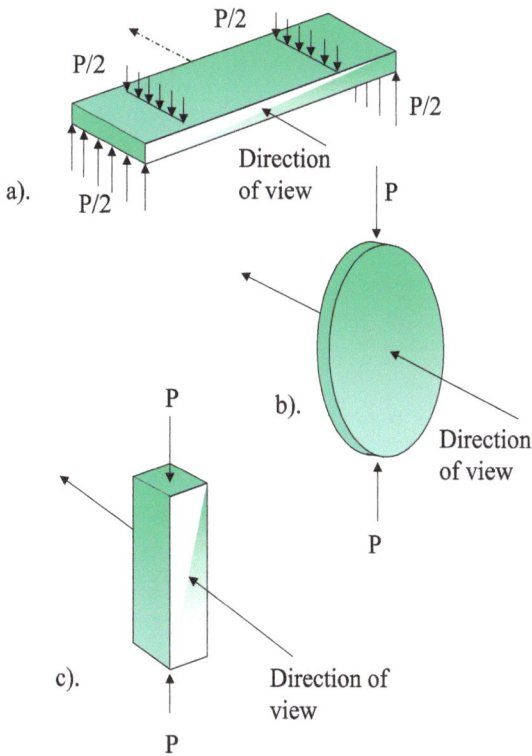

Figure 2.18. Test specimens for stress optical coefficient measurement

2.4.3   Evaluation of stress optical coefficient

The value of the stress optical coefficient used in Equations (2.3)–(2.5) for the glass type being analysed may be unknown. If this is the case there are several approaches which can be used to obtain a value for this property. These all entail loading a simple test specimen, made of the glass in question, in a standard fashion and measuring the retardation at a position on the specimen, as the load is changed. For glass the most straight-forward test geometries for such a calibration are:

(i) a beam in pure bending (Figure 2.18(a)),

(ii) circular disc under diametral compression (the so-called

Brazilian disc test (Figure 2.18(b))

(iii) a uniform compression test. (Figure 2.18(c)).

For the beam test:

$$C = \frac{I\delta}{Myt} \qquad (2.6)$$

where I is the second moment of area of the beam (mm^4), M is the bending moment, (N mm), y is the distance from the neutral axis (mm) to the measurement position, b is the optical path length in the beam (mm) and δ is the retardation (=$N\lambda$) (nm). As the stress field varies with distance from the neutral axis it is possible to obtain several readings from just one load level (especially if a compensation method is being used to measure the relative retardations; see Section 3.3).

The Brazilian disc test can also yield several readings from just one load level as the stress distribution is well known (see, for example, Hondros [5]) however, in this case, it is easier to examine the centre of the disc and take data at several load levels. At the centre of the disc:

$$C = \frac{\pi R \delta}{4Pt} \qquad (2.7)$$

where R is the radius of the disc (mm), P is the applied load (N) and t is the disc thickness (mm).

The uniform compression test is one in which a rectangular parallelepiped of glass is compressed axially and the retardation noted as the load is changed. In this instance, for a sample having a square cross section of side, b (mm) subjected to a load, P (N):

$$C = b\frac{\Delta\delta}{\Delta P} \qquad (2.8)$$

where $\Delta\delta/\Delta P$ is the slope of the graph of δ plotted against P.

3. Basic techniques

3.1 Polariscopes

As mentioned in Section 2.4.1, a polariscope is an instrument for observing and analysing photoelastic fringes appearing within a stressed specimen. At the simplest level, it consists of a light source, a polariser and an analyser, between which the specimen to be analysed is placed. The light source may be such that the light from it can be passed through a lens to provide collimated light which will fall on the sample at normal incidence (i.e. at 90° to the sample surface) or, alternatively, a diffuser may be incorporated into the lamp housing to produce diffuse illumination of the specimen. A mercury vapour lamp is an example of the former light source, while the diffuse light source may simply consist of a bank of tungsten or sodium bulbs with a diffuser plate in front. In general, larger fields of view can be obtained with diffuse light sources and hence, if glass products such as windscreens or television faceplates have to be examined, this type of light source is preferred.

Polariscopes may use either plane or circular polarisation. The optical arrangement for the first case is termed a plane polariscope, as described previously and shown in Figure 2.12. This polariscope enables both isoclinics and isochromatics to be observed. A circular polariscope consists of a light source, polariser and analyser, together with two quarter-wave plates, as illustrated in Figure 2.15. In operation, the incident light passes through the polariser, then the first quarter-wave plate, followed by the specimen. After the light has passed through the specimen, it passes through the second quarter-wave plate and then the analyser. Only the isochromatic fringes are observed in this case.

There are occasions when it is necessary to be quite precise about the level of retardation present in a particular region of interest in a glass being examined in a polariscope. However, compared with other photoelastic materials, all glasses have relatively low values of stress optical coefficient (see Table 2.1) which means that, for a given stress difference the magnitude of the retardations, and hence the number of fringes, is generally quite small. To obtain larger and more easily measurable retardations or to evaluate fractional fringe orders arising between two integer fringes, which can otherwise be distinguished visually by their colours in white light, compensators are often used.

3.2 Compensators

Compensators are optical devices which may be used to produce a controlled retardation when placed in series with the glass being measured. The level of retardation arising from the compensator may be varied continuously, thereby facilitating measurement of the retardations in the glass under examination. There are a number of compensation devices and techniques available and some of the more common ones used in the glass industry are described in the following paragraphs.

3.2.1 Wedge

One of the simplest forms of compensator to use is a plastic wedge having frozen-in residual stresses. A wedge may be produced by manufacturing a rectangular bar from epoxy resin which is subsequently taken through a thermal cycle to freeze a uniform tensile stress into the bar. The bar is then cut on a diagonal to produce a wedge. The stress within the wedge is uniform but, as the thickness of the wedge varies along its length, the retardation also varies along its length from zero at the tip to some higher order at the opposite end. Thus when a wedge is viewed between crossed polarisers in white light a series of coloured parallel fringes (orientated across the width of the wedge) are seen. Figure 3.1(a) shows a wedge viewed in a circular polariscope from which the coloured integer fringes

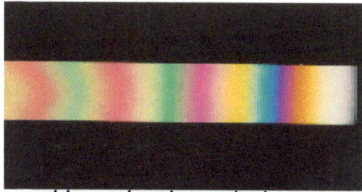

(a) Wedge viewed in a circular polariscope using white light.

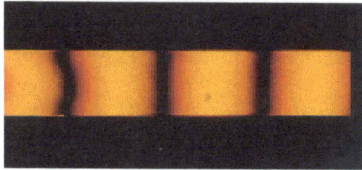

(b) Wedge viewed in a circular polariscope using monochromatic light.

Figure 3.1. Wedge viewed in white and monochromatic light

can be seen, in accordance with Table 2.2. The zero order fringe, which is always black, is located at the tip of the wedge while the first order fringe is centred on the deep blue/deep red boundary. The second order fringe centre corresponds with the deep green/deep pink boundary and, as the fringe orders increase, the intensity of the colours decrease becoming more indistinct at higher orders. When a wedge is viewed in monochromatic light, a series of black fringes is seen, Figure 3.1(b) and the wedge can be graduated into convenient fractional distances between these integer black fringes. With monochromatic light higher order fringes can be seen more clearly.

The tensile wedge compensator is most suited to assessing retardations at the free edges of glass products such as glass plates, windscreens and some containers. In addition, since the principal stress perpendicular to the edge is necessarily zero, the wedge is particularly useful for determining whether the non-zero stresses parallel to the edges of such glass components are tensile or compressive. This may be done as follows:

(a) Place the glass to be assessed in a white light circular polariscope.

(a) Position wedge normal to the edge

(b) Look for the formation of a black fringe at the edge

(c) Stresses at the edge

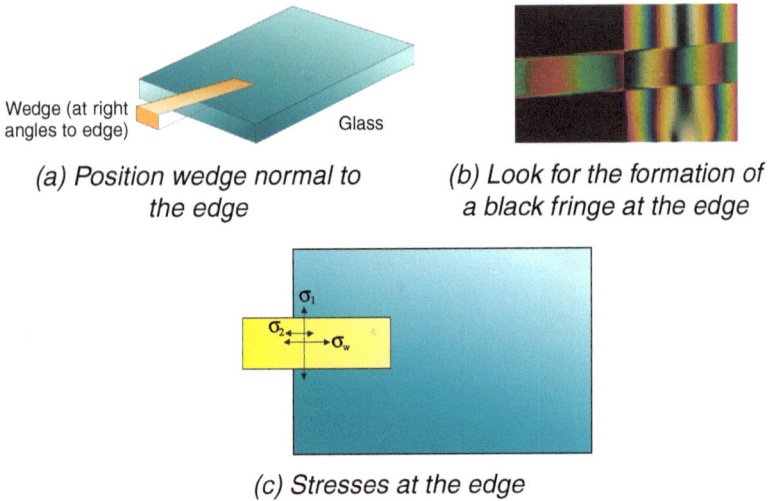

Figure 3.2. Use of a wedge to check for tensile stresses at the edge of a glass

(b) Hold the wedge against the glass at right angles to the edge and move the wedge in and out in the direction of the main axis of the wedge, Figure 3.2(a).

(c) As (b) is carried out look for a black fringe to appear on the edge, such as that shown in Figure 3.2(b). If this does occur, then the edge stress is tensile, as explained below.

With reference to Figure 3.2(c), photoelastic analysis yields the following. The light that passes through the glass and the wedge is affected by the stresses in both. Thus, taking the stress in the wedge to be σ_w, the stress parallel to the edge of the glass to be σ_1 and the stress normal to the edge to be σ_2, the total stress at right angles to the edge is $(\sigma_2+\sigma_w)=\sigma_w$ since $\sigma_2=0$. Therefore, using the stress optic relationship (Equation (2.3)), gives:

$$(\sigma_1 - \sigma_w) = \frac{N\lambda}{Ct} \qquad (3.1)$$

Since a black fringe is seen, $N=0$ thus,

$$\sigma_1 = \sigma_w \qquad (3.2)$$

BASIC OPTICAL STRESS MEASUREMENT IN GLASS

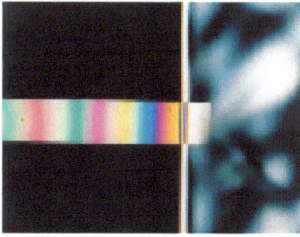

(a) With the wedge in this orientation, the fringe order at the edge increases

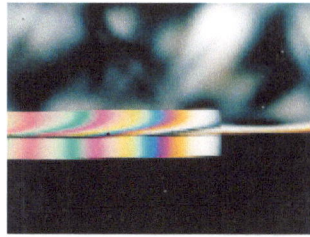

(b) With the wedge parallel to the edge, the fringe order at the edge becomes zero

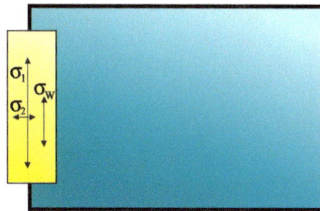

(c) Stresses present at the edge

Figure 3.3. Use of a wedge to check for compressive stresses at the edge of a glass

and since σ_w is tensile, σ_1 must also be tensile.

Alternatively, if it is not possible to obtain a black fringe with the wedge orientated in this way, Figure 3.3(a), then the wedge should be turned so that it is parallel to the edge of the glass, Figure 3.3(b).

Again the wedge should be moved in the direction of its long axis.

If a black fringe appears on the edge, as shown in Figure 3.3(b), then the stress at the edge is compressive, as the following shows.

In this case (see Figure 3.3(c)), the total stress parallel to the edge is $(\sigma_1+\sigma_w)$ whilst the stress perpendicular to the edge $\sigma_2=0$. Thus using Equation (2.3) for the wedge and the glass,

$$(\sigma_1 + \sigma_w) = \frac{N\lambda}{Ct} \qquad (3.3)$$

(a) Without glass   (b) With glass

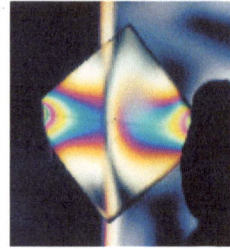

Figure 3.4. Application of a polyurethane square to the determination of the sign of the stress at the edge of a glass

Again $N=0$ and, therefore

$$\sigma_1 = -\sigma_w \qquad (3.4)$$

and since σ_w is tensile, the negative sign indicates that σ_1 must be compressive.

A measure of the fringe order at the edge can be made by placing the wedge on the glass, as described previously, and noting where the black fringe appears in relation to the nearest lowest integer fringe value. The fringe order is then given by the lowest integer value plus the fractional part between this value and the next integer above it. Graduation of the wedge between integer fringes, as mentioned previously, assists in assessing the fractional fringe order.

3.2.2 Polyurethane square

A similar approach can be used with a polyurethane square, such as that shown in Figure 3.4. It will be noticed, when it is squeezed in plane, the polyurethane square is very sensitive photoelastically, however, on this occasion, since squeezing provides a reference stress which is compressive, the orientations relative to the glass edge for identification of tension and compression, are opposite to those given above for the tensile wedge, i.e. the stress in the glass in Figure 3.4(b) is compressive.

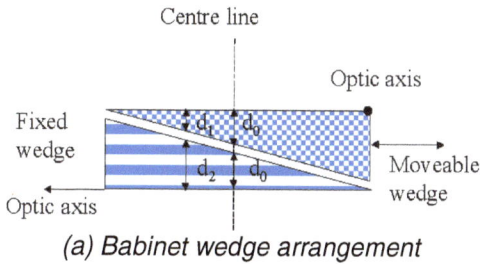

(a) Babinet wedge arrangement

2 1 0 1 2

(b) Babinet fringes

Figure 3.5. Babinet compensator wedges and fringe pattern

3.2.3   Babinet compensator

A Babinet compensator consists of two independent quartz, or calcite, wedges which are orientated as shown in Figure 3.5(a). The wedges have the same angle which may be between 2° and 3°, dependent upon the number of fringes required. The optic axes of the two wedges are perpendicular to each other and one wedge is fixed whereas the other can be moved by rotation of a calibrated thimble, similar to that on a micrometer. Any light ray passing through the compensator travels a distance d_1 in the first wedge and a distance d_2 in the second wedge. The retardation generated is constant along lines running across the width of the compensator, as the wedge thicknesses are constant along such lines. Thus a series of parallel fringes are generated across the width of a Babinet compensator. Along the line where $d_1=d_2=d_0$ there is zero retardation and, a black fringe will be seen. On either side of this position, the fringe order increases progressively. This is shown in Figure 3.5(b).

The Babinet compensator is used by incorporating the compensator in the polariscope between the glass to be examined

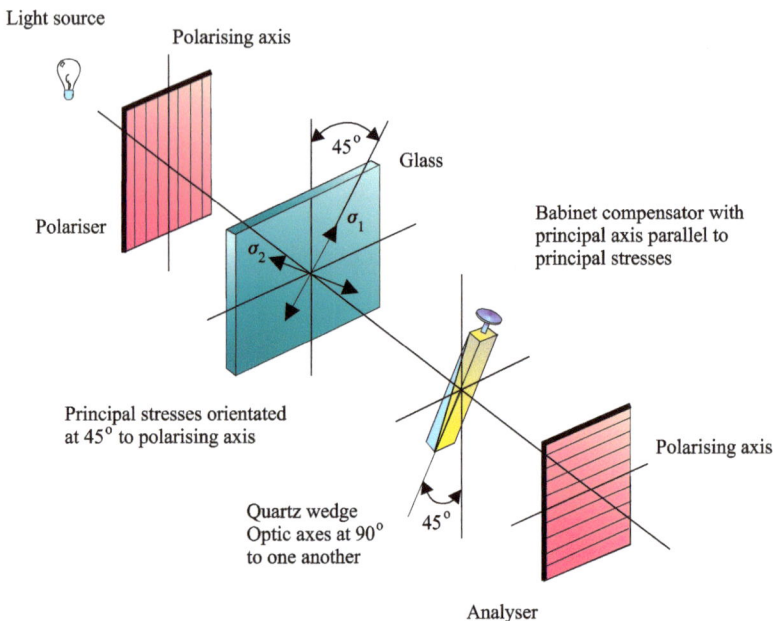

Figure 3.6. Arrangement of polariscope when using
Babinet compensator

and the analyser as shown in Figure 3.6.

The initial reading of, say, the zero order fringe is noted by centring this on the cross hair, without the glass in place. The glass is then inserted into the polariscope and the compensator adjusted by rotating the graduated thimble until the zero order fringe returns to its initial position. From a knowledge of the calibration constant for the compensator, the net thimble rotation can be converted to retardation and then to principal stress differences using the approach which follows.

Calibration of the Babinet compensator can be done using white or monochromatic light. If white light is used, it is done by centring the zero order fringe on the cross hair in the instrument and noting the reading on the micrometer, then rotating the thimble until the centre of the first order fringe is coincident with the cross hair and again noting the reading on the barrel. The difference between the two readings (Δr) is then the wedge movement necessary for a change in retardation of one wavelength. This is known as the calibration constant for the Babinet compensator. In the case of white light, the wavelength is taken to be that of the so called *tint of passage* (λ=577 nm) which corresponds to the sharp change from red to blue in the first order fringe.

To determine the stress difference using the Babinet compensator, Equation (2.3) can again be used, thus:

$$(\sigma_1 - \sigma_2) = \frac{N\lambda}{Ct} \tag{3.5}$$

where N is net thimble rotation for glass being measured divided by Δr, t is glass thickness or path length of light through the glass, λ is 577 nm, and C the stress optical coefficient.

3.2.4 Babinet-Soleil compensator

A Babinet-Soleil compensator also consists of two wedges except in this case, they are cut with their optic axes parallel to each other. However, in addition, a separate quartz plate of uniform thickness is placed in parallel with the wedges, the optic axis of the plate being perpendicular to the optic axes of the wedges, as shown in Figure 3.7.

The essential difference between this compensator and the Babinet compensator is that instead of seeing separate fringes when viewed in a polariscope, the retardation seen in the Babinet-Soleil field is uniform. The retardation can be altered by rotating

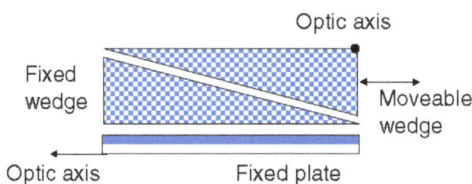

Figure 3.7. Arrangement of the optical elements of a Babinet-Soleil compensator

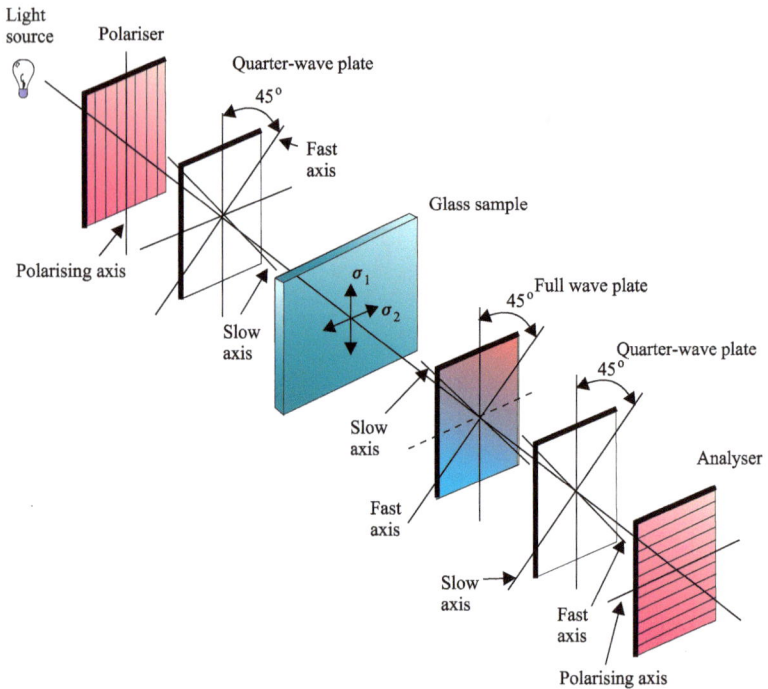

*Figure 3.8. Optical arrangement required
when a tint plate is used*

the thimble which, as before, slides one wedge over the other. Thus the Babinet-Soleil compensator is effectively a variable retardation plate.

The precision in retardation measurement of both the Babinet and Babinet-Soleil compensators is ~0.01λ.

3.2.5   Tint plate

When examining some forms of glassware, a detailed analysis of the stresses present may not be necessary, it being sufficient to know that the level of retardation falls within acceptable limits determined from practical experience. Such an assessment can be done by incorporating a full-wave plate into a circular polariscope with a white light as the source of illumination. The arrangement of the optical elements is shown in Figure 3.8.

This arrangement is particularly useful when the magnitude of the retardation to be evaluated is small. The full-wave plate is made from the same material as a quarter-wave plate (e.g. mica) and behaves in a similar way, except the thickness of the plate is such that a phase difference of one wavelength is developed between the light vibrating in the two orthogonal principal planes through the thickness of the plate.

When a full-wave, or tint, plate is introduced into a circular polariscope set at 45° to the polariser and analyser axes as shown in Figure 3.8, then viewed through the analyser, it appears a magenta colour as shown in Figure 3.9(a). With the addition

(a) Tint plate in a circular polariscope

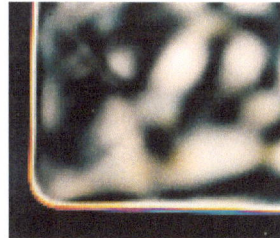

(b) Toughened glass in a circular polariscope

(c) Toughened glass with tint plate in a circular polariscope

Figure 3.9. Application of a tint plate to the analysis of stresses in a toughened glass

Table 3.1. *Effect of full-wave tint plate on the fringe colour for different retardations*

Retardation (nm)	Colour
+325	Yellow
+275	Yellow/Green
+200	Green
+145	Blue/Green
+115	Blue
0	Magenta
−25	Red
−130	Orange
−200	Bright Yellow
−260	Yellow
−310	White
−565	Black

of a stressed glass sample, having the retardation shown in Figure 3.9(b), the field then seen through the analyser, appears as shown in Figure 3.9(c).

Experience in the use of full-wave plates enables the sign of the stresses (i.e. tension or compression) to be determined from the colour change which takes place with the tension effectively increasing the overall retardation and compression reducing it. Table 3.1 gives an indication of this effect.

3.2.6 Standard strain discs

The factors which influence the colours seen when examining a glass article in a polariscope incorporating a tint plate, or full-wave plate, are as follows:

(a) the retardation in the glassware itself which may be attributed to the residual stresses in the glass, as well as the glass thickness along the line of sight (See Equations (2.3)–(2.5)).

(b) the nature of the stress in the glass, i.e. whether it is tensile or compressive.

(c) the glass type from which the article is made and its associated stress optical coefficient.

(d) the glass colour – it is sometimes not possible to obtain the

Figure 3.10. Standard set of strain discs

normally recognisable colours due to absorption of specific wavelengths by the glass itself.

(e) the retardation of the tint plate.

With regard to (e) it should be noted that a small variation in the tint plate retardation, say ±10 nm, can cause noticeable differences in the colours seen when a glass sample is examined in the polariscope. Thus, unless tint plates are standardised on a specific wavelength, to within fairly tight tolerances, a 'red' colour, for example, may indicate quite different glass retardations when one tint plate is used compared with another.

One way of overcoming this difficulty is to compare the colours seen when examining glasses in a plane polariscope with discs of known retardations, referred to as strain discs, such as those introduced in the 1930s by the Glass Container Association of America. The Standard Strain Discs consist of 2·5 mm thick glass discs heat treated in such a way as to develop a retardation of 23±2 nm at a distance of 6 mm from the perimeter. The discs are mounted in sets of five giving possible retardations, by combination, of 23, 46, 69, 92 and 115 nm. These retardation values and the number of discs were selected after a study of the range of likely retardations to be encountered in practice, particularly in the glass container industry. A set of Standard Strain Discs produced by the then Department of Glass Technology of the University of Sheffield is shown in Figure 3.10.

In use, the glass to be examined is placed in the polariscope with the strain discs positioned adjacent to the glass. An assessment of the retardation in the glass is made by comparing the colours seen by looking at the glass and the strain discs, the latter being altered, by the addition or subtraction of discs, until the closest match between the two is obtained. For any given glass product it is possible to stipulate, from practical experience, that colours corresponding to, say, two discs or less is indicative of an adequate standard of annealing. Anything above such a limit would require the glassware to be re-annealed.

It is worth noting that the colours seen in any part of the glass product are a result of the stresses lying in a plane normal to the direction of view. Stresses along the line of sight do not affect the colour. For example, in examining containers by viewing through the sides, the colours seen are a result of the stresses along the length and across the face of the container, integrated through the overall glass thickness. It would be necessary to examine a ring section of the container in order to detect any stress through the thickness of the container wall. The colours seen when viewing through the sides indicate the average stress difference through the total glass thickness in the direction of viewing. If a bottle, for example, had equal, but opposite, stresses on the two sides, the colour seen would be black. At first glance this may suggest that the bottle was stress free but, as stated previously, such a conclusion could be incorrect since the effect seen is an integrated one through the total thickness (See also Section 2.4.2).

When examining containers, such as jars, bottles or drinking glasses, by holding them in a polariscope, the shape of such products provides a distorted view due to reflection and refraction of the light at the curved surfaces. Thus, where more detailed analysis of these products is required optical distortion can be avoided by immersing the glass to be studied in a rectangular glass tank containing a clear fluid having the same refractive index as the glass container (see Section 3.4).

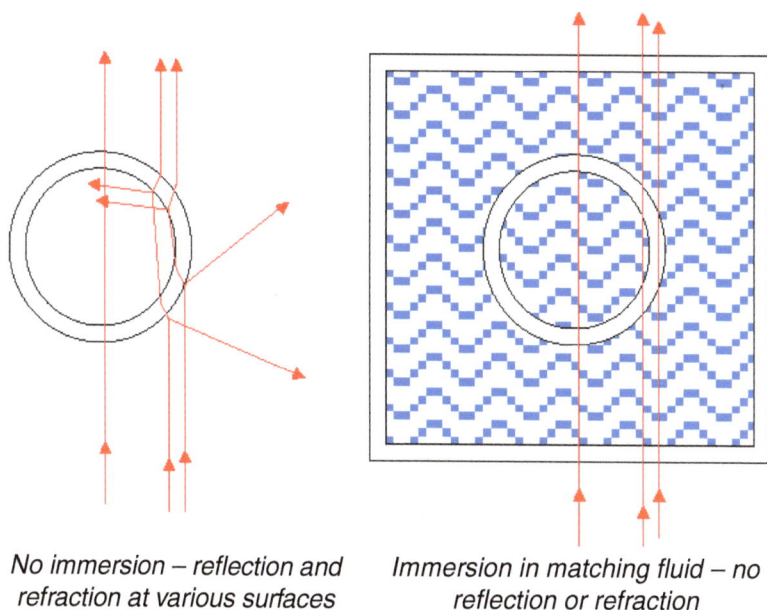

No immersion – reflection and refraction at various surfaces

Immersion in matching fluid – no reflection or refraction

Figure 3.11. Immersion of a glass in a transparent tank containing a liquid having the same refractive index as the glass

This avoids undesirable refraction and reflections as shown in Figure 3.11.

3.2.7   Berek compensator

This is a compensator often used in conjunction with a polarising microscope, principally due to its very compact construction (Hallimond [6]). It consists of a calcite plate, approximately 0·1 mm thick, cut perpendicular to the optic axis and supported on a short spindle. The plate is tilted about the horizontal position by the spindle, the angle of rotation, which is limited to approximately ±20°, being registered on the graduated spindle knob. The scale is calibrated using monochromatic light with the compensator located between a polariser and an analyser.

The results from the calibration lie on a curve for which the theoretical form can be calculated and which is supplied with the instrument.

In use, the compensator is positioned between the glass being examined and the analyser of a plane polariscope. The compensator scale is set to zero and the polariser and analyser rotated together to bring an isoclinic to the position of measurement. When this is done, the polariser and analyser are rotated together a further 45°, after which, measurement of retardation in the glass can be made by means of the compensator. In this case, compensation is made by bringing the dark band to the centre of the field and restricting the aperture to a central beam.

3.3 Goniometric compensation

It is possible to determine fractional fringes and to make an accurate measurement of the total retardation at a given location on a glass by using the optical elements of the polariscope without the need for additional equipment, such as has been discussed so far. Into this category come the Sénarmont and Tardy methods of compensation. Of the two, the former is more often used in the glass industry, particularly because it uses only one quarter-wave plate, whereas the second method requires two. In the early days of photoelasticity this would also have been the choice, principally because the method of producing quarter-wave plates, which entailed splitting mica sheets to the required thickness, was not particularly accurate and to obtain a matched pair was even more difficult. Hence the preference for the method which makes use of only one quarter-wave plate. Many of the standard references on photoelasticity give the mathematical background to both the Sénarmont and Tardy methods, (for example, Kuske & Robertson [3]), and so these will not be repeated here. Rather the setting up procedure for each will be given and these are as follows.

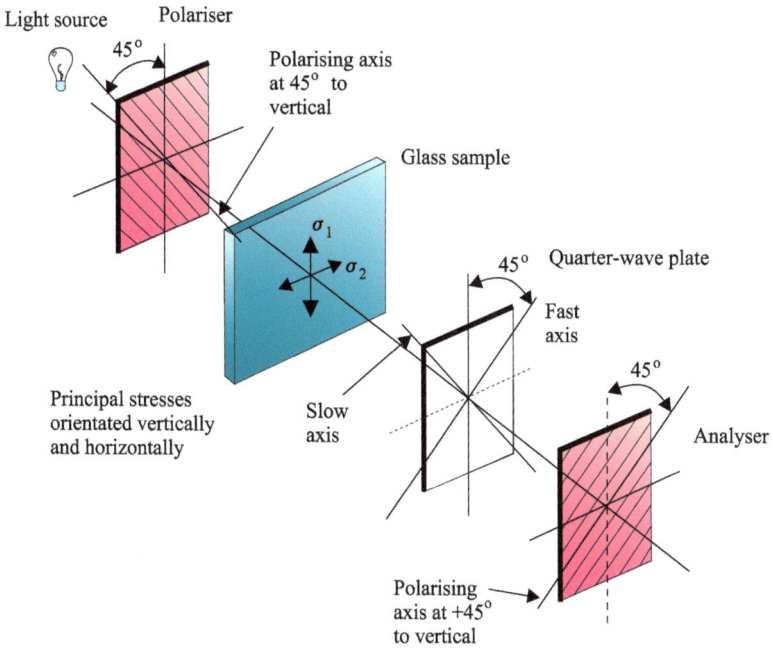

Figure 3.12. Polariscope arrangement for Sénarmont compensation

3.3.1 Sénarmont compensation

(a) With the quarter-wave plate removed from the field, cross the polariser and analyser (i.e. produce a dark field as seen viewing passed the glass) and rotate these two elements together (i.e. coupled) until an isoclinic passes through the point of interest in the glass. The axes of polarisation of the polariser and analyser are now aligned with the directions of the principal stresses σ_1 and σ_2 at this position, thereby enabling the orientation of the principal stresses to be determined relative to a given datum, say, for example, the axis of polarisation of the analyser or polariser.

(b) Rotate the polariser and analyser together through a further 45°. By so doing, the isoclinics are moved such that they are then at their furthermost position from the point of interest.

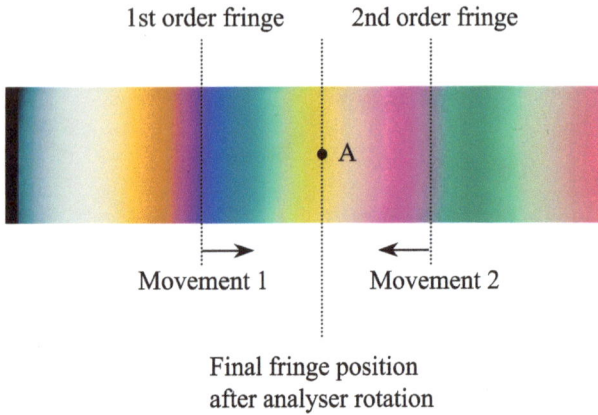

Figure 3.13. Diagrammatic representation of fringe movement during Sénarmont compensation

(c) Uncouple the polariser and analyser, to enable the analyser to be rotated separately.

(d) Insert the analyser quarter-wave plate and rotate to give a dark field. The arrangement of the polariscope at this stage, which is that immediately prior to measurement, is shown in Figure 3.12.

(e) By inspection, determine the fringe order in the region of the point of interest (see Figure 3.13).

(f) Up to this point white light can be used, but the measurement of the fractional fringe order which now follows is best made using monochromatic light. Rotate the analyser on its own in a clockwise direction to move one of the bounding fringes to the point of measurement and note the angular rotation of the analyser which has been made. The interpretation of the measurement may be made with reference to Figure 3.13.

If in the initial set-up the first order fringe is moved to the point by an angular rotation θ_1 (in degrees) of the analyser, then the fringe order at A is:

$$N = \left(1 + \frac{\theta_1}{180}\right) \qquad (3.6)$$

Alternatively, by rotating the analyser in an anti-clockwise direction, the second order fringe may be moved to the point of interest by a rotation of θ_2 (in degrees.) In this case the fringe order at A is:

$$N = \left(2 - \frac{\theta_2}{180}\right)$$ (3.7)

Irrespective of whichever direction of rotation is carried out the value of N should be the same and thus both measurements should be made to check that the results are consistent.

It should be noted that clockwise rotation of the analyser will not always move the lower order fringe to the point of measurement or vice-versa. The direction of rotation is dependant upon the alignment of the fast axis of the quarter-wave plate and the polarising axis of the analyser. For example, if the quarter-wave plate is rotated through 90°, the direction of rotation of the analyser during compensation will be opposite to the original direction in order to bring about the same fringe movement.

In a similar way, the direction of rotation of the analyser to move the lower fringe order to the point of interest changes if this point changes from being in a compression zone to a tensile one, everything else remaining the same.

3.3.2 Tardy compensation

(a) The first step is the same as for the Sénarmont method.
(b) Disconnect the polariser and analyser.
(c) Insert both quarter-wave plates to give a circular polariscope, thereby eliminating the isoclinics. The polariscope arrangement at this stage is shown in Figure 2.15.
(d) By inspection, determine the fringe order in the region of the point of measurement.
(e) Select monochromatic light and proceed as described in (f) for the Sénarmont method. The interpretation of fringe order is as previously described.

Table 3.2. Refractive indices of various glasses (after Aben & Guillemet [4])

Glass	Refractive Index
Quartz glass	1·45–1·46
Optical fibres	1·46–1·49
Optical glass	1·46–1·90
Crowns	1·47–1·65
Container glass	1·49–1·53
Flat glass	1·51–1·53
Lanthanum glasses	1·80–2·00

3.3.3 Friedel compensator

This compensator may be considered to be a circular polariscope, with compensation being undertaken using the Tardy method.

3.4 Immersion fluids

To obtain photoelastic measurements where the angle of incidence of the light striking the surface of the glass is such that refraction may take place, then it is necessary to immerse the glass under investigation in a fluid which has a refractive index very closely matched to that of the glass in question; the difference in refractive indices should not be more than 0·002. This fluid is known as an immersion fluid and it must be contained within a bath that has transparent plane parallel sides which are stress free, otherwise the bath itself will contribute to the observed photoelastic fringes. The sides of the bath should be perpendicular to the light beam which is used to illuminate the specimen under investigation. In practice it is also necessary for the fluid to have a low viscosity to minimise the possibility of bubble entrapment within the fluid. Such bubbles lead to unwanted diffraction and interference effects, thereby hindering the interpretation of photoelastic data generated when using immersion fluids.

Tables 3.2 and 3.3 give the refractive indices of various glasses and immersion fluids, respectively. The dependence of refractive

Table 3.3. Refractive indices of various immersion fluids (after Aben & Guillemet [4])

Fluid	Refractive index
Ethyl alcohol	1·362
Olive oil	1·460
Silicone oils	1·43–1·61
Sunflower oil	1·472
Turpentine	1·473
Glycerine	1·474
Furfuryl alcohol	1·486
Dibutyl phtalate	1·489
Toluol	1·497
Dimethylphtalate	1·516
Cereclor	1·493
Liquid paraffin	1·468

indices on wavelength means that the accurate matching of immersion fluids to the glasses can really only be guaranteed for monochromatic light. In addition, it can be seen from Tables 3.2 and 3.3 that obtaining a suitably accurate refractive index match for a particular glass with any one fluid could be difficult. It may thus be necessary to mix two (or even three) fluids with different refractive indices, one higher and the other lower than that of the glass specimen. It should be noted, however, that not all the fluids can be successfully mixed. The relative quantities of the fluids in the mixture can be varied until a suitable refractive index match is obtained. A simple test for the match is to view a ruler held at approximately 45° behind the immersion bath containing the specimen. If the refractive indices are matched the ruler will appear unbroken when viewed from the other side of the bath. If the ruler appears broken and the image of the ruler seen through the glass sample is higher than the image seen through the bath alone then some more of the lower refractive index fluid is required. If the break in the image is the opposite way round, then more of the higher refractive index fluid needs to be added to the mixture.

As the above method of mixing involves trial and error, equations have been developed for calculating the amounts of two different fluids of known refractive index (n_1 and n_2, respectively) and density (ρ_1 and ρ_2, respectively) which have to be mixed to give the desired refractive index, n, (see Smolik & Bellow [7] and Singh [8]). Assuming that there is no volume change on mixing, then the volumes of fluids 1 and 2 (V_1 and V_2, respectively) that are required to achieve the desired refractive index are given by:

$$V_1 = \frac{100(n_2 - n)}{(n-1)(\rho_2 - \rho_1) - \rho_2(n_1 - 1) + \rho_1(n_2 - 1)} \qquad (3.8)$$

$$V_2 = \frac{100 - V_1\rho_1}{\rho_2} \qquad (3.9)$$

It should be noted that refractive indices can vary with both wavelength and temperature. It may therefore be necessary to take either or both of these factors into account when using the above formulae.

4. Flat glass

Residual stresses, arising from either the annealing or toughening process to which the glasses have been subjected, can vary in a three-dimensional manner. This makes absolute stress measurement using transmission photoelasticity difficult, as has already been described in Section 2.4.2. To simplify matters, the overall stress distribution may be considered to consist of *area* (or *membrane*) and *thickness stress* components, with *bending stresses* occurring occasionally. The membrane stresses are assumed to be uniform through the thickness, but vary over the area of the glass and are due mainly to temperature differences which occur between the edges and centre of the glass as it cools through the transformation zone. Temperature differences, which occur at the same time, between the glass surfaces and mid-thickness are responsible for the formation of the thickness stresses, which vary in a near parabolic manner for both annealed and thermally toughened glasses. The area and thickness stress distributions are shown diagrammatically in Figure 4.1.

In the paragraphs that follow, methods are described which enable the area and thickness stresses to be measured.

4.1 Float glass: on-line measurements

4.1.1 Manual measurements

Area stresses in float glass can be measured using any of the compensation methods described in Sections 3.2 and 3.3 but, generally, measurements are made using either the tint plate (Section 3.2.5) or Sénarmont method (Section 3.3.1). In order to manually measure retardations on-line, a full ribbon-width light box with a polarising sheet on top, is positioned below the glass. Since the principal stresses at the edges of a glass ribbon are oriented parallel and perpendicular to the edges then, to avoid isoclinics masking the isochromatics, the polarising axis

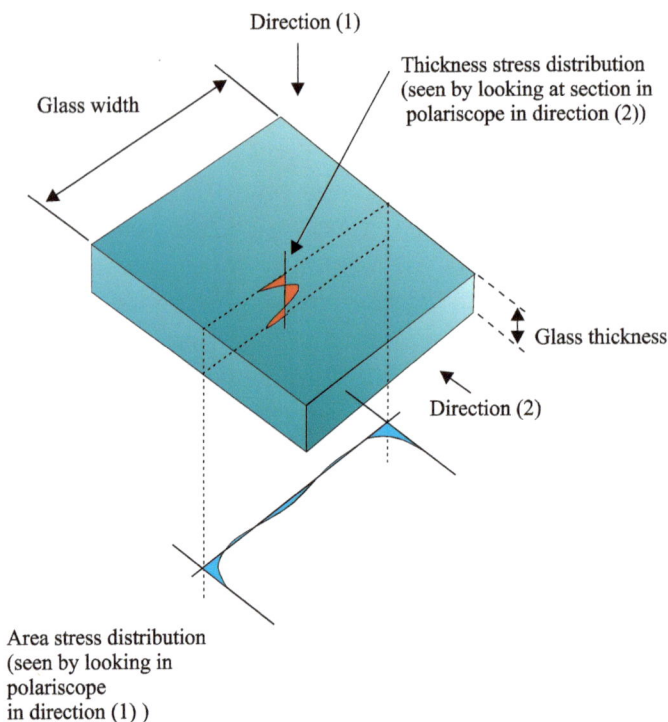

Direction (1)

Glass width

Thickness stress distribution
(seen by looking at section in
polariscope in direction (2))

Glass thickness

Direction (2)

Area stress distribution
(seen by looking in
polariscope
in direction (1))

Figure 4.1. Area and thickness stresses

of the polariser is set at 45° to the edge for these measurements,
in accordance with Section 3.3.1. A hand-held viewer with a
quarter-wave plate and analyser, correctly oriented relative to
the polariser, see Figure 3.12, is used to examine the glass from
above the ribbon by the Sénarmont method. A separate viewing port in the viewer houses a tint plate for colour recognition
of the retardations. An illustration of the hand-held viewer is
shown in Figure 4.2.

4.1.2   Automatic strain viewers

Various designs of automatic strain viewers have been used.
Some have been based on the principle of a moveable quartz

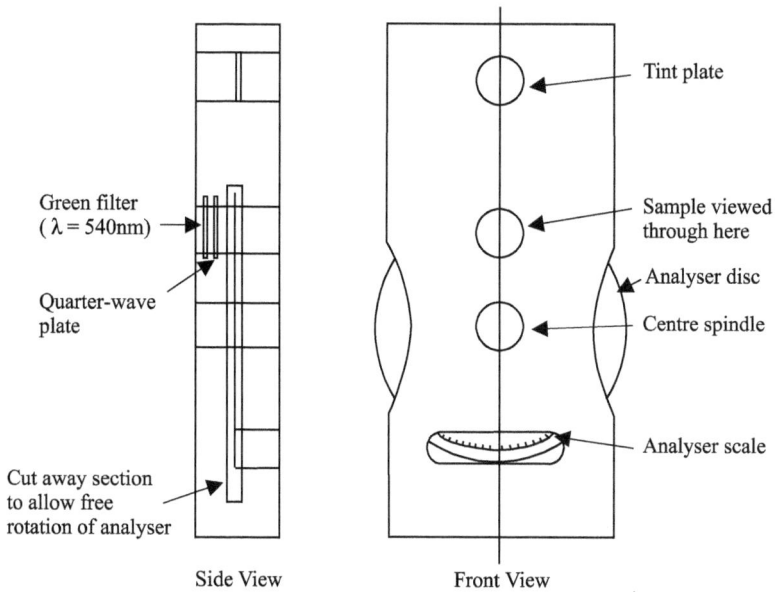

Figure 4.2. *Hand held viewer for measurement of area stress in float glass on-line*

wedge, the position of which is controlled by two photodetectors, one on each side of the visible fringe. A feed-back control loop between the detectors and the motor driving the wedge ensures that the wedge is positioned such as to maintain a balanced output from the detectors. By so doing, the retardation in the wedge compensates for the retardation in the glass at each measurement position. Calibration of the system enables the relationship between wedge movement and retardations to be established for a given wavelength of light. The principle of this method of compensation is basically that previously described in Section 3.2.1.

More recent automatic strain viewers use a variant of the Sénarmont technique in which a rotating analyser is used as a means of measuring the retardation in the glass ribbon. Such strain viewers can be located upstream where the glass is hot

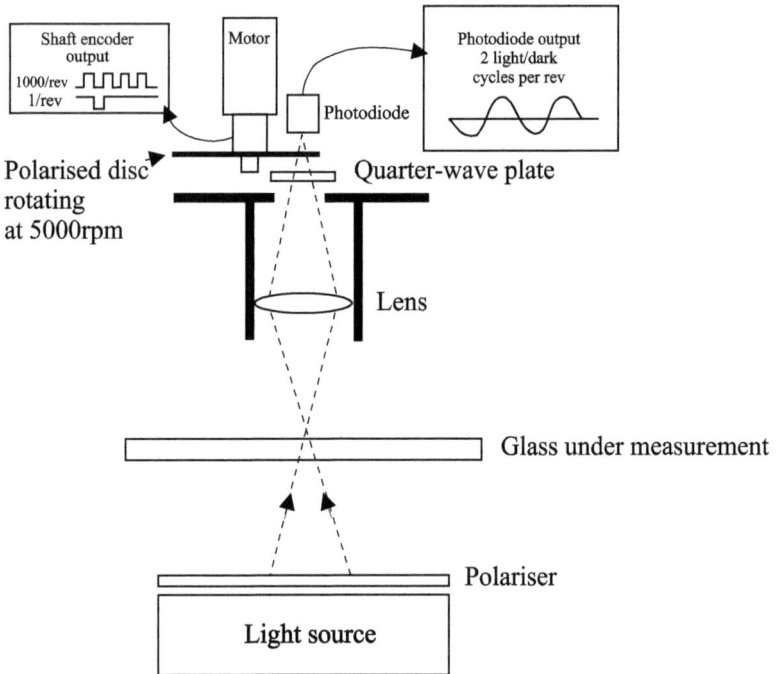

Figure 4.3. Schematic diagram of an automatic
on-line strain viewer

and then, by simultaneously recording the glass temperature and the retardation across the width of the ribbon at the position of measurement, the area stress in the glass at room temperature can be predicted using software on a computer linked to the apparatus.

A schematic diagram of the optical/electronic arrangement of such a strain viewer is shown in Figure 4.3.

The light source comprises two 20 Watt green fluorescent tubes connected in series to a 40 Watt high frequency (32 kHz) electronic ballast, while the optical elements are arranged as shown in Figure 3.12 with their polarising axes at 45° to the ribbon edge. In operation, the analyser rotates continuously at ~5000 rpm and the photodiode positioned over the disc gener-

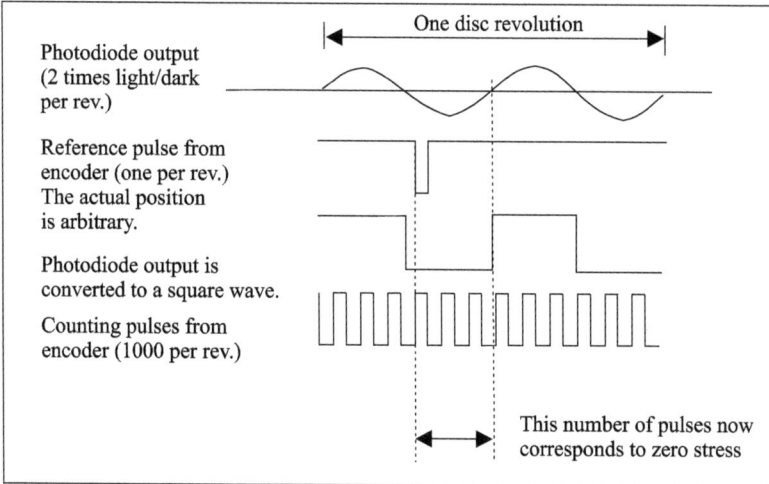

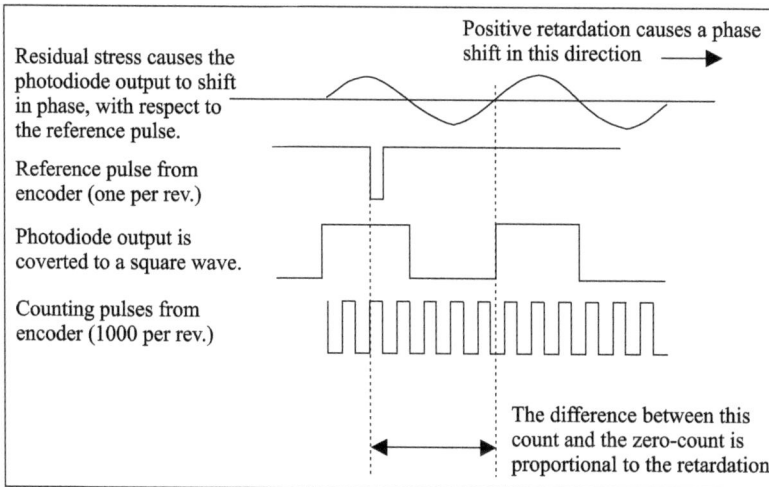

Figure 4.4. Electronic output (top) without glass and (bottom) with glass present

ates an output proportional to the intensity of light it sees. The combination of the fixed polariser and the rotating analyser causes the photodiode to see two dark/light cycles per revolution. Thus the photodiode outputs a sinusoidal wave at a frequency twice that of the motor speed. A shaft encoder fitted to the motor generates two sets of pulses, a reference signal (one pulse per revolution) and a counting signal (1000 pulses per revolution).

With no glass present, the number of counting pulses between the reference pulse and one of the dark/light transitions is called the zero-stress count and this is depicted in Figure 4.4(a). When residually stressed glass is introduced, a shift in the positions of light and dark develops as a consequence of the retardation provided by the glass. The phase shift of the sine wave is thus proportional to the retardation and hence the area stress in the glass, as shown in Figure 4.4(b). In operation, the number of pulses between the reference pulse and the new position of the dark/light transition is counted. The difference between this count and the zero-stress count is proportional to the residual stress at the measurement location.

The output of the shaft encoder is mechanically coupled to the disc position. This enables a degree of immunity to motor speed variations, as well as providing a facility to manually change the position of the reference pulse. As the phase shift is measured relative to the reference pulse, its actual position is arbitrary, thus allowing the instrument to have variable ranges such as, −270/+270 nm, −100/+440 nm or −440/+100 nm.

4.2 Annealed glass – area stress measurement

Although, in general, any of the methods of compensation described earlier can be used to make measurement of area stresses in annealed glass, there are occasions when transmission measurements cannot be made. This may be due to obscuration bands, such as those on car windscreens, or when accurate measurements of retardation have to be made on glass components where the magnitudes are very low. In such circumstances, the following methods may be used.

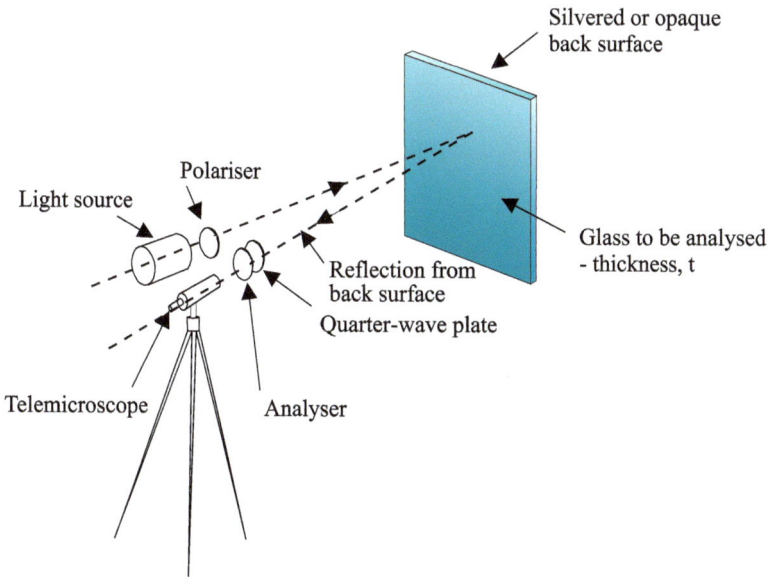

Figure 4.5. General arrangement for reflection photoelasticity

4.2.1 Reflection photoelasticity

Reflection photoelasticity is undertaken by projecting polarised light at the glass and examining the reflected light from the back surface by means of an analyser. An advantage of this method is that since the light passes through the glass twice the retardation is effectively doubled. In the case of clear glass, a reflecting surface or layer can be used, whereas in the case of windscreens with obscuration bands, the band itself will serve as a means of reflecting the transmitted light. However, due to absorption by the band, the reflected light level is low and fringe clarity is reduced. The arrangement of the apparatus is shown in Figure 4.5.

The interpretation of the fringe pattern seen when reflection photoelasticity is used is similar to that for normal transmission except that, in reflection, the path length through the glass is

doubled, thus,

$$(\sigma_1 - \sigma_2) = \frac{N\lambda}{2Ct} \qquad (4.1)$$

or, for a given stress difference,

$$N = \frac{2Ct(\sigma_1 - \sigma_2)}{\lambda} \qquad (4.2)$$

which is twice the value obtained in transmission for the same stress difference.

4.2.2 Pockels cell

The principle of this method centres around the Pockels cell, which is an electrically variable crystal retarder (See Collett [9]). In a transverse Pockels cell, retardation is produced by the direct application of an electrical voltage to the faces of the crystal. As the retardation developed in this way is a linear function of the applied voltage, the magnitude of the retardation can easily be

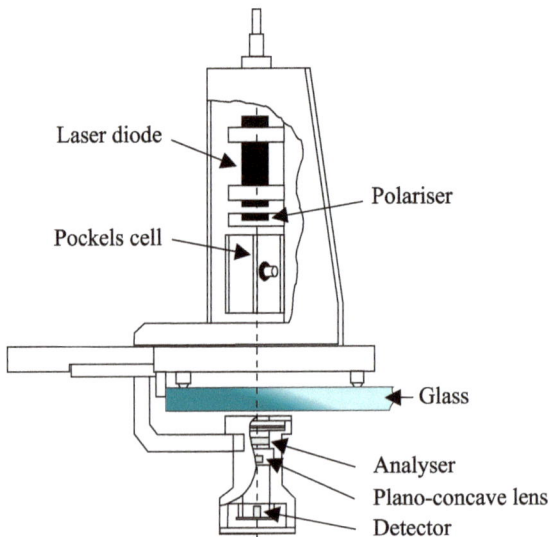

Figure 4.6. Edge stress meter incorporating a Pockels cell

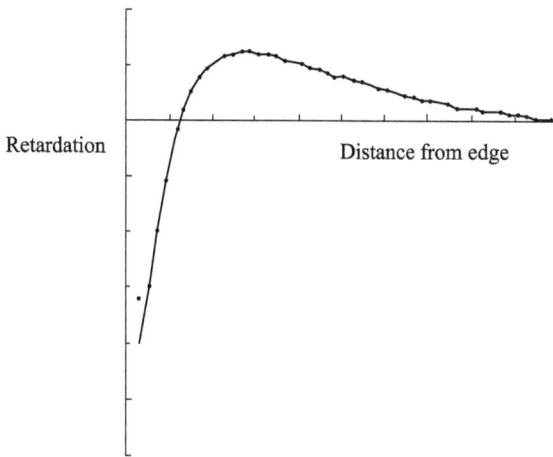

Figure 4.7. Typical variation of retardation with distance from edge for a windscreen

altered, a quarter-wave retardation change taking only a few nanoseconds.

An instrument has been developed, based on the Pockels cell, which is operator-independent and is capable of measuring reliably and repeatably, low levels of retardation in the edge region of a glass, such as a windscreen. The instrument is light and portable and can be used in a plant environment, with minimum set-up time. The instrument electronics are so arranged as to give information regarding the retardation profile for some distance in from the edge of the glass. In addition, the retardations at all measurement positions can be output to a PC for further data processing and storage for future reference.

The apparatus used for this purpose is shown diagrammatically in Figure 4.6 and comprises a laser diode as the light source and a polariser, positioned immediately before the Pockels cell. Once through the Pockels cell the light then passes through the glass, through an analyser, and diverging lens into a detector. A feed-back loop between the detector and the Pockels cell ensures that, by maintaining zero intensity at

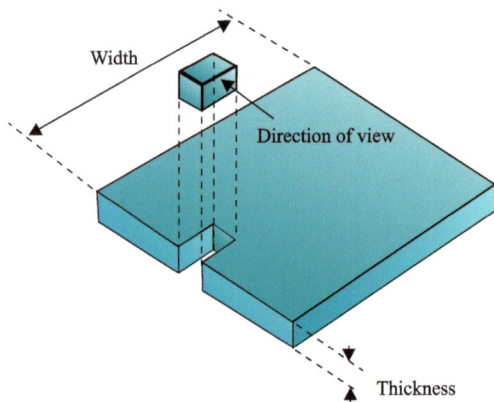

Figure 4.8. Sample for thickness stress evaluation and direction of view for Babinet measurement

the detector, the correct voltage is applied to the Pockels cell to develop a retardation which compensates exactly for the retardation in the glass. A typical plot of retardation against distance from the glass edge is given in Figure 4.7.

4.3 Annealed glass – thickness stress measurement

Thickness stresses in annealed glass are normally measured by the Babinet compensator. The stresses are measured using a small sample cut from the main plate, as shown in Figure 4.8.

The width of the sample for such measurements is nominally 25–50 mm, dependent upon the level of residual stress. Examination of the sample is made in the direction shown in Figure 4.8, using the optical arrangement shown in Figure 4.9 and yields a retardation profile such as that shown in Figure 4.10. Measurement of the stresses is carried out using the procedure outlined in Section 3.2.3. In this case, it is assumed that the stress perpendicular to the surfaces of the original plate is zero through the complete glass thickness, and hence the stress difference from this analysis reduces to the stress normal to the

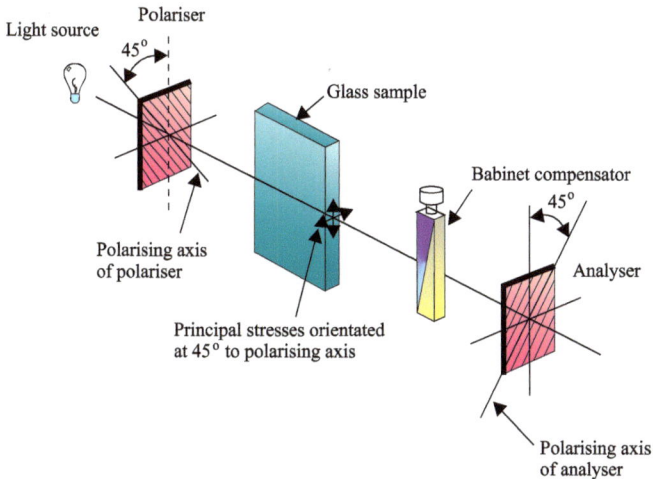

Figure 4.9. Optical arrangement for the measurement of thickness stress by means of a Babinet compensator

edge of the original plate wherever the measurement is made through the thickness. However, it should be noted that the action of cutting the small sample from the parent plate will result in some release and redistribution of residual stresses and hence measurements made in this way will not truly reflect the stresses in the uncut plate.

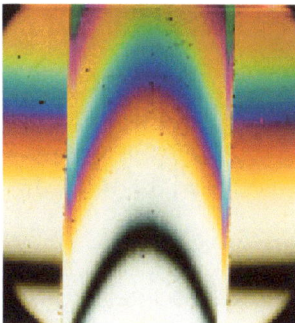

Figure 4.10. Babinet fringe pattern for thickness stress

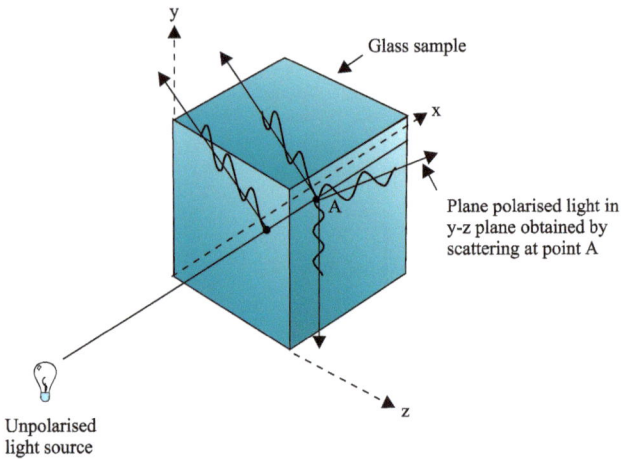

Figure 4.11. Scattering of light within glass sample

4.4 Toughened glass – area stress measurement

Any of the compensation methods discussed in Sections 3.2 and 3.3 can be used for the measurement of area stresses in toughened glass, but the Sénarmont method is possibly the preferred option (See Section 3.3.1), though wedge compensators used near to the glass edges can also be effective.

4.5 Toughened glass – thickness stress measurement

As toughened glass cannot be cut to provide samples similar to those described in Section 4.3, an alternative approach must be adopted for the measurement of thickness stress in toughened glass. The method commonly chosen for this is the scattered light technique (See Holister [10] and Dally & Riley [11]).

4.5.1 Scattered light method

This is a method based on the scattering characteristics of a wave as it passes through the glass, a concept which is illustrated by reference to Figure 4.11.

A wave of ordinary light propagating in the x direction through the sample would normally, for transmission work, be examined in this direction. However, frequently 100% transmission does not occur and light scattering along the length of the beam takes place. This scattered light can be considered as secondary vibrations excited by the main wave and which propagate radially outwards from this source. That is, for a main wave of ordinary light propagating in the x direction, the vibrations associated with the scattered light will all lie in planes normal to the x axis such that, when the scattered light is viewed along any ray perpendicular to the x direction it will be seen to be plane polarised. From this it may be assumed that there is an infinite number of such scattering sources each producing polarised light in a direction radially outward from the main source. It is possible to use this effect in two ways:

(a) using ordinary light and, in terms of conventional photoelasticity, allowing the scattering medium to act as the polariser. The scattered light is observed by means of an external analyser.

(b) using polarised light as the main source and allowing the scattering effect to serve as the analyser. The resulting fringe pattern may then be observed by means of a telemicroscope or scanned by a photomultiplier device.

As the latter method enables the information to be obtained with greater ease than the former, it is more commonly used and will be described in some detail.

On entering the stressed sample, the beam of plane-polarised light, say from a laser source, splits into two planes corresponding to the principal stresses σ_1 and σ_2 at entry, as shown diagrammatically in Figure 4.12.

The two components travel with different velocities along the principal planes such that the resultant light vector generally traces an elliptical path in the plane of the wavefront. Where the two components are exactly in phase with one another the ellipse becomes a straight line at 45° to the two components. In such a

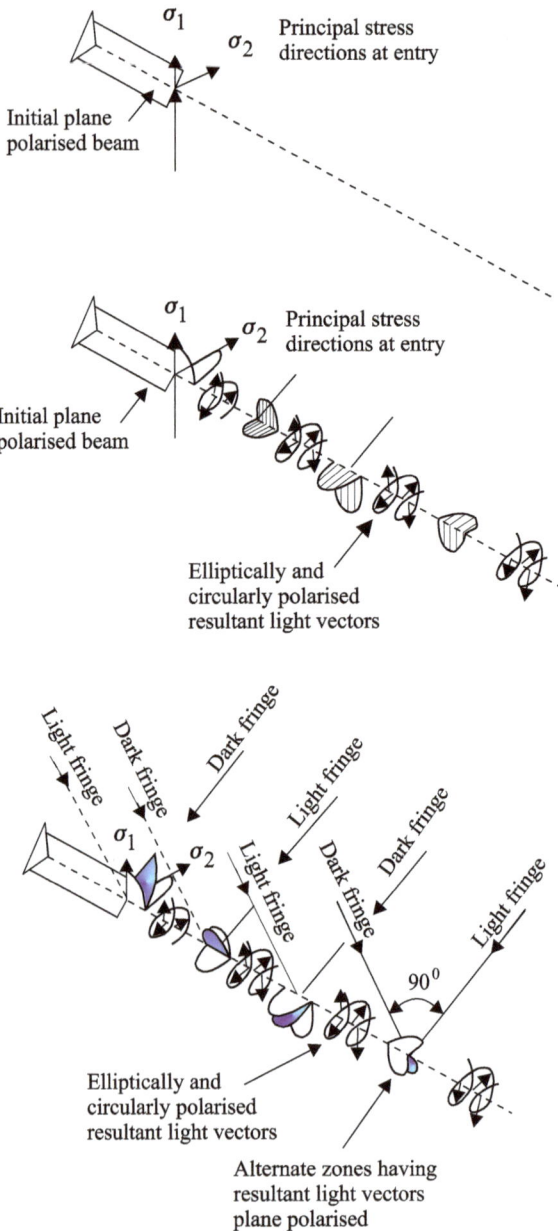

Figure 4.12. Fringe development in scattered light

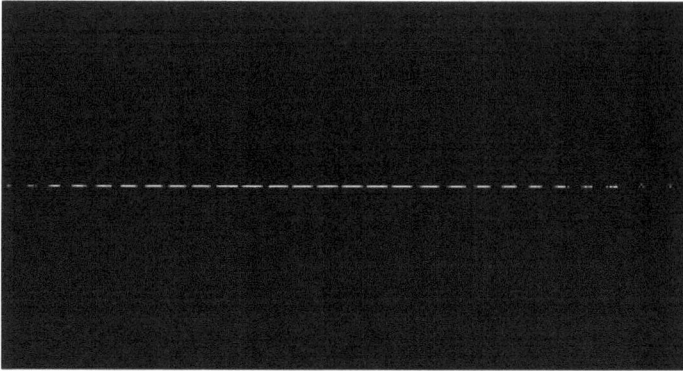

Figure 4.13. Scattered light fringe pattern

region the light is plane polarised. On observation from a position outside the sample and along a line of sight parallel to this resultant vibration, no scattered light will be observed. However, if the two components arrive at the point exactly opposite in phase, the ellipse will become a line at 90° to the previous one and will be normal to the line of sight. Hence a maximum intensity of scattered light will be seen. As the light propagates through the sample and the two components progressively alternate in phase, variations in intensity of the scattered light along the initial beam direction can be seen from certain observation positions outside the sample. This variation in intensity constitutes interference fringes as shown in Figure 4.13.

For example, in the measurement of residual stresses in a flat glass plate, maximum fringe clarity and sharpness can be found when the plane of polarisation at entry to the sample is at 45° to the plane of the glass and the direction of viewing of the beam as it travels through the glass is also at 45° to the glass surface. The method is ideally suited to measuring the maximum tensile stress mid-way through the glass thickness and is not normally used to measure surface compressive stresses due to the rapidly changing stress state near to the surface, particularly when considered in relation to the laser beam diameter.

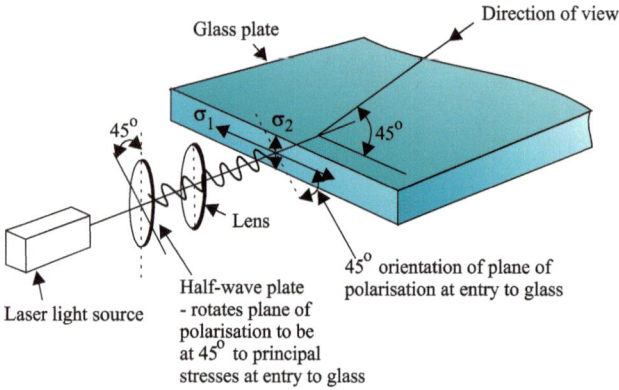

Figure 4.14. Optical arrangement for scattered light measurements in thin glass plates

Ideally the laser beam diameter should be as small as possible which generally means focusing the beam of ~1 mm diameter down to an acceptable size in the region of interest in the sample, dependent upon the glass thickness and the corresponding rate of change of stress through the thickness. Helium neon or argon lasers, having power outputs of 10–20 mW, are suitable for this application and, in order to have some degree of control over the orientation of the plane of polarisation for optimum fringe clarity, a half-wave plate should be inserted between the laser and the glass to be measured. Thus the optical arrangement is as shown in Figure 4.14.

Referring to Figure 4.15, the interpretation of the fringe pattern in terms of stress may be obtained. Thus:

$$(\sigma_1 - \sigma_2) = \frac{\Delta N}{\Delta S} \cdot \frac{\lambda}{C} \tag{4.3}$$

where σ_1 and σ_2 are the principal stresses lying in a plane at right angles to the direction of propagation of the light; $\Delta N/\Delta S$ is the slope of the fringe number/distance graph at the position of interest in the glass and λ and C are as defined in Section 2.4.2.

In glass plates it is reasonable to say that the stress at the

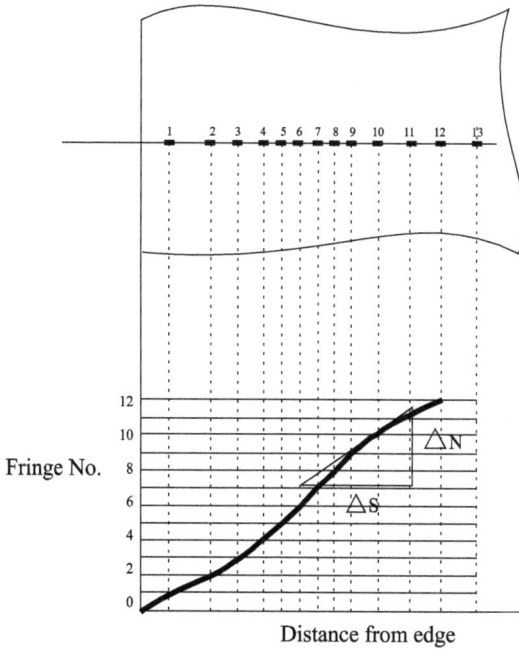

Figure 4.15. Interpretation of fringe pattern in terms of stress

mid-thickness position in a direction normal to the surface, i.e. σ_2, is zero (this is correct actually on the free surfaces) which then enables the other stress lying in the plane of the plate to be obtained directly, i.e.

$$\sigma_1 = \frac{\Delta N}{\Delta S} \cdot \frac{\lambda}{C} \qquad (4.4)$$

4.6 Surface stress measurement

As a means of assessing the effectiveness of a toughening process, whether it be thermal or chemical, surface stress measurements are often made. The instruments used for this purpose are generally those available commercially and these fall into two groups, namely, those based on the principle of surface refractometry (Ansevin [12], Guillemet & Acloque [13], Kishii [14]) and those based on surface polarimetry (Guillemet & Acloque [15], Redner [16]). A brief account of each now follows.

4.6.1 Surface refractometry

The magnitude of the surface stress is directly proportional to the birefringence of the surface for light polarised parallel and perpendicular to the surface. This may be expressed by an alternative form of the stress optic law to that given previously, namely,

$$\sigma = C(\eta_1 - \eta_2) \qquad (4.5)$$

where σ is the stress in the surface perpendicular to the plane of incidence; C is the stress optic coefficient for the glass; η_1 and η_2 are the refractive indices for light vibrating perpendicular and parallel to the plane of incidence, respectively.

Refractive index values measured with a critical angle refractometer are, in general, accurate to a few parts in the fourth decimal place. As this is insufficient for stress determination, refractive index differences are measured which increases the accuracy by at least an order of magnitude. Hence the name of one surface stress measuring instrument, the differential surface refractometer (DSR) (See DSR Manual [17] or Ansevin [12]).

The principle of operation of the DSR is that light is passed through a prism resting on the glass to be measured, such that total internal reflection can take place at the glass surface. The refractive index of the prism is higher than that of the glass being measured and a fluid of intermediate index is used between the prism and the glass to provide optical contact. The prism, as shown in Figure 4.16(a), is divided into two parts, entrance and exit, separated by an opaque divider which prevents directly reflected light from entering the telescope. Thus only light which has entered the surface of the glass at the critical angles

(a) DSR prism arrangement

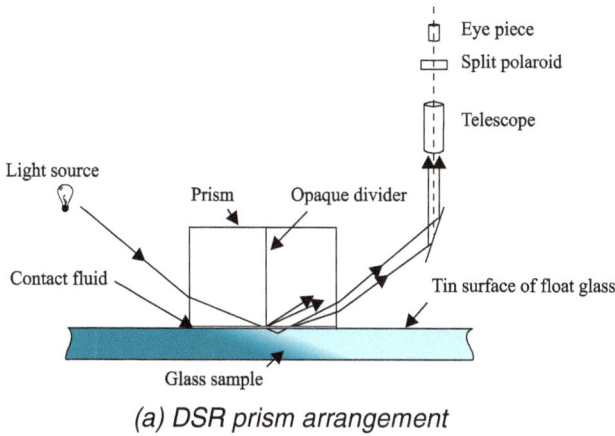

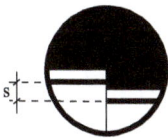

(b) Fringe pattern for
thermally toughened
glass

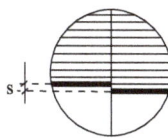

(c) Fringe pattern
for chemically
toughened glass

Figure 4.16. Optical arrangement for differential surface
refractometer and diagrammatic representation of fringes for
thermally and chemically toughened glass

corresponding to η_1 and η_2, travelled through the glass close
to the surface in the exit prism region and then re-entered the
prism at the critical angles, is allowed to reach the telescope.
This light does not re-enter the prism at a single point, but rather
enters gradually over the entire face of the exit portion of the
prism to form two parallel bundles of rays corresponding to
each polarised component. These bundles of light are focused
in the telescope into two bright lines, one formed from the light
polarised parallel to the sample surface and the other from the
light polarised perpendicular to the sample surface. The separation of the two lines is directly proportional to the surface

stress lying in the plane of the glass surface at right angles to the axis of the instrument. Provision of a micrometer eye-piece permits measurement of the spacing, s, between corresponding lines on each side of the split field (Figure 4.16(b) and (c)). The stress on the surface in a direction normal to the principal axis of the prism is given by:

$$\sigma = ks \qquad (4.6)$$

where k is a constant.

It should be noted that the constant linking the fringe spacing and the surface stress is not necessarily the same for all instruments and all glass types and hence a calibration should be carried out to determine k for each glass type. In the case of chemically toughened glass, it is not sufficient to assess k by basing it on a bulk-glass value of the stress optical coefficient, since the chemical composition and the refractive index of the glass close to the surfaces is not the same as that of the bulk glass.

The calibration can be carried out using a beam made from the relevant glass and toughened, either thermally or chemically, whichever is appropriate. The beam, with strain gauge rosettes applied to confirm the analytical assessment of the applied stress, should be loaded in four-point bending. The instrument to be calibrated must be placed on the top surface, transverse to the main axis of the beam; such an orientation ensures the measurement of the principal stress due to bending.

Figure 4.17 shows an arrangement of a four-point bending apparatus which is capable of applying tensile or compressive bending stresses to the top surface of the beam, since the calibration ought to be done under both conditions.

The procedure recommended is to take a zero reading from the instrument with no bending applied, and then to make successive readings as the bending stress is increased. The calibration constant is given by the slope of the stress/DSR reading (corrected for the initial value) graph. Due to the vari-

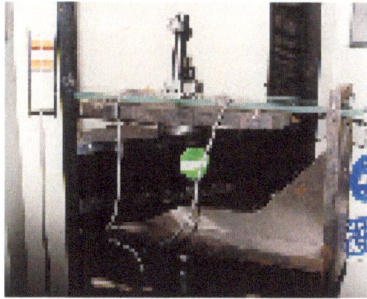

(a) Glass beam in tension

(b) Glass beam in compression

Figure 4.17. Arrangement of four-point bending apparatus for DSR
calibration

ability in readings which have been noticed when making this
evaluation, it is recommended that the procedure is repeated
a number of times, in both tension and compression, to ensure
a reliable value for the constant.

This method of surface stress measurement has some limitations, including the need for the glass surface to be flat, as well
as demanding precision in the measurement of the effects of
small differences in refractive index. Also, the method operates
most effectively where there is a surface layer which acts as a
wave guide. Thus the instrument works well on the tin side of

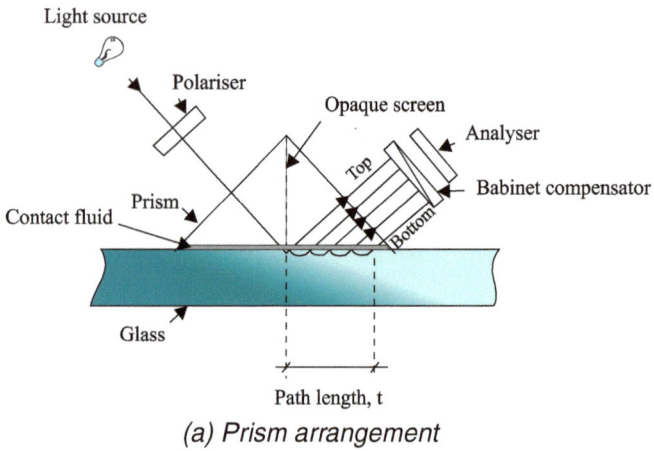

(a) Prism arrangement

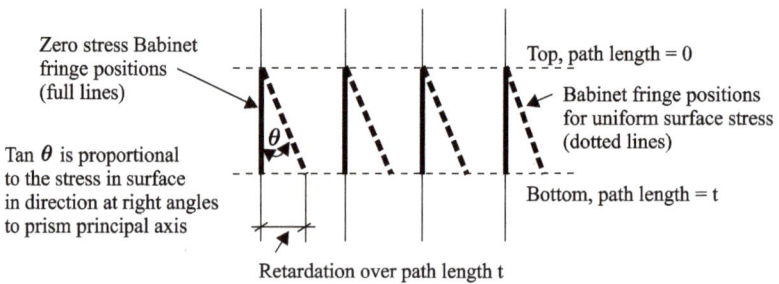

(b) Diagrammatic representation of Babinet fringes

Figure 4.18. Optical arrangements and fringe pattern for surface polarimetry

thermally toughened float glass and on the non-tin surface of chemically toughened float glass. The fringe patterns seen when examining thermally and chemically toughened glasses are shown diagrammatically in Figure 4.16(b) and (c), respectively. Normally only two fringes, one on each side of the split field of view, are seen when thermally toughened glass is examined, while multiple fringes are seen when chemically toughened glass is studied. In the case of the latter, only the bottom two fringes are used to evaluate the stress but the other fringes can be used to obtain an indication of the stress profile normal to the glass surface, as described by Aben & Guillemet [4].

4.6.2 Surface polarimetry

The principle of operation of this method of surface stress analysis is also based on the incident light striking the glass at the critical angle. The method makes use of the wave guide effect provided by the tin-rich layer of float glass, whereby the refractive index variations in this layer are such that the incident ray gives rise to several successive curved refracted rays. These turn upwards towards the prism, as shown in Figure 4.18(a), and then pass through a quartz wedge or Babinet compensator for analysis.

With no stress present, the Babinet fringes are in the positions shown by the full lines in Figure 4.18(b). When uniform stresses are present, the fringes rotate to the positions shown by the dotted lines such that the total retardation over the path length, t, is given by δ. Calibration of the instrument enables a link to be established between the tangent of the angle through which the fringes are rotated, relative to their start position, and the surface stresses. It should be noted that compressive surface stresses will cause the fringes to rotate in one direction while tensile surface stress will cause the fringes to rotate in the opposite direction. Thus discrimination between tensile and compressive stresses can be made using this method.

One of the main differences between this form of surface stress assessment and the surface refractometry approach is that the measurement is dependant upon the path length, t, such that:

$$\sigma = \delta/Ct \qquad (4.7)$$

This being so, the measurement is more easily made, with an increase in measurement sensitivity estimated to be 5 to 10 times that of differential surface refractometry. However, this measurement method is restricted to the tin surface of float glass and is not suitable for chemically toughened glass.

4.7 Laminated glass

The measurement of residual area stresses in laminated glass can be made by any of the methods described previously in Sections 3.2 and 3.3. While the interlayer (normally PVB) is birefringent, it does not contribute significantly to the overall retardation to be measured. Also, it should be recognised that there is the possibility that bending stresses will also be present, in addition to area and thickness stresses, due to slight mismatches in curvature of glass pairs. Such stresses cannot be measured by transmission techniques as they integrate to zero along the direction of view and hence recourse must be made to surface polarimetry specially adapted for low levels of residual stress.

5. Container glass

Detailed evaluation of residual stresses in container glass from photoelastic measurements is more involved than the equivalent evaluation of residual stresses in flat glass due to the more complex shapes of containers. However, simple quality assurance assessments may be undertaken, without attempting to evaluate the absolute magnitude of the residual stresses, by examining the glassware in a polariscope and using either a full-wave plate as a tint plate in the manner described in Section 3.2.5 or a standard strain disc as described in Section 3.2.6.

Where more detailed analyses are required on containers, alternative methods such as those given in the following paragraphs may be used.

5.1 Glass containers – ring sections

As mentioned in Section 3.2.6, ring sections can be cut from glass containers to establish the stress variation through the glass thickness. Such sections are also useful for assessing the homogeneity of the glass as indicated by the amount of cord (or striae or ream) present. The method adopted for such an analysis is to obtain a ring section approximately 10 mm in height from the upper third of the container sidewall. The ring section is then immersed in a dish containing a matching fluid and examined in a direction parallel to the longitudinal axis of the container in a polarising microscope (Figure 5.1) with a full-wave tint plate oriented at 45° to the polariser and analyser axes (see Section 3.2.5) in the field of view. The ring is therefore viewed at positions parallel and normal to the tint plate axes.

Upon examining the ring section in this way consideration must be given to the following:
(a) the amount of cord present, i.e. fine threads, broad bands
(b) the intensity of colour in the cords
(c) whether the cord is tensile or compressive

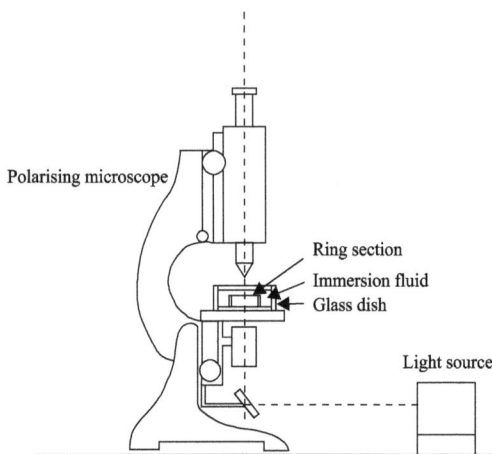

Figure 5.1. Polarising microscope arrangement for examining ring sections

(d) the location of the cord in the section, i.e. outer or inner surface.

An assessment is made having regard for the above points and the results referred to a cord rating system chart (see Table 5.1) which links the level of tension, as given by the colours observed, (see Table 3.1) and the extent of the inhomogeneity. In Table 5.1, the letters A-D refer to the amount of cord present regardless of whether stresses are mild or severe, while the numbers 1-5 give an indication of the magnitude of the tension present. The X or XX record whether tension appears on the outer surface for more or less than 10% of the circumference. A limit line exists on one side of which combinations are acceptable and so too are the glasses, while combinations on the other side of the line are deemed to be unacceptable.

5.2 Glass rods and tubes

During the production of glass rods and tubes, residual stresses are produced by the interaction between thermal contraction, elasticity at low temperature, viscoelastic behaviour at higher

Table 5.1. Ring section assessment (for 10 mm deep rings)

	Tensile stress up to: 5·5 N/mm² Blue	Tensile stress up to: 6·2–6·9 N/mm² Blue/Green	Tensile stress up to: 9–10·3 N/mm² Green	Tensile stress up to: 12·4–13·8 N/mm² Green/Yellow	Tensile stress up to: 13·8 N/mm² Yellow/ Orange/Red
No inhomogeneity or cord	A	-	-	-	-
Traces of cord appearing as fine threads	B1	B2	B3	B4 X / XX	B5
One or more fine cords each less than 10% of glass thickness	C1	C2	C3 X / XX	C4	C5
Wider bands of cord each 10% or more of glass thickness	D1	D2 X / XX	D3	D4	D5

N.B. The stress values given above correspond to the retardations given in Table 3.1 and a stress optical coefficient of ~2·1 Brewsters

temperature and temperature gradients caused by cooling. These stresses (axial and hoop) are compressive on the outer surface and tensile at the centre of the rod or the inner surface of the tube. The compressive stresses on the outer surface are desirable from the point of view of the glass strength as they inhibit microcrack growth, but they are undesirable, to a certain extent, from a cutting standpoint. Thus the measurement and control of these residual stresses is important.

A possible method for the measurement of residual stresses involves cutting a thin disc or ring from the end of the rod or tube and then examining it in a way similar to the ring sections described for containers. This is a destructive method and cannot always be applied, particularly when the residual stresses are very high or when the rod or tube has small lateral dimensions. Further, the ring cutting process changes the residual stresses in the disc or ring and, in addition, viewing the discs in this way does not permit the direct measurement of the axial stresses.

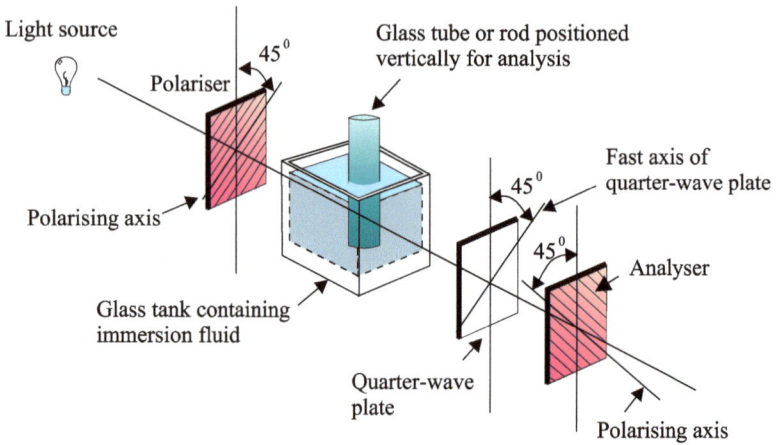

Figure 5.2. Arrangement of apparatus for measurement of rods or tubes

An alternative method has been described by Sutton [18], Ritland [19] and Read [20]. This entails viewing the rod or tube in a tangential direction, as shown in Figure 5.2, whilst the rod or tube is immersed in a tank containing a liquid which has the same refractive index as the glass being examined.

As shown in Figure 5.2, the polariser and analyser are set at ±45° to the vertical direction which corresponds to the longitudinal axis of the tube or rod since the axial, hoop and radial stresses are the principal stresses and these lie parallel and perpendicular to this direction. Measurement of retardation is made tangentially by viewing through a telemicroscope at incremental distances from the outer perimeter to the centre, in the case of rods, or to the inner surface, in the case of tubes. Compensation is undertaken using the Sénarmont technique described in Section 3.3.1 previously. Once the retardation has been measured the principal stresses can be evaluated, using the approach described by Sutton [18], Ritland [19] and Read [20]. However, this approach is rather complex, and thus the following simplified methods which give dependable results for both rods and tubes have been developed.

5.2.1 Rods

Measurement in rods generally reveal that the retardation is zero at a distance of $R/2$, from the centre of the rod where R is the radius of the rod. Also, the stress distributions are usually parabolic in form, with the value of the axial stress (σ_z) at the centre of the rod being tensile and equal, but opposite in sign, to that on the perimeter. Furthermore, the axial and hoop stresses (σ_θ) are equal in magnitude and sign on the perimeter while the radial stress (σ_r) is zero on the perimeter and equal to the hoop stress at the centre. Thus:

$$\sigma_{z\,\text{centre}} = \sigma_{z\,\text{surface}} \qquad (5.1)$$

$$\sigma_{z\,\text{surface}} = \sigma_{\theta\,\text{surface}} \qquad (5.2)$$

$$\sigma_{r\,\text{surface}} = 0 \qquad (5.3)$$

$$\sigma_{r\,\text{centre}} = \sigma_{\theta\,\text{ centre}} \qquad (5.4)$$

If the value of retardation at the centre of the rod (N_{rc}) is measured then a simplified relationship enables the axial stress to be calculated as follows:

$$\tilde{A}_{z\,\text{centre}} = \frac{1{\cdot}5 N_{rc}\lambda}{CR} \qquad (5.5)$$

where λ is the wavelength of light used in the analysis (nm); C is the stress optic coefficient for the glass (Brewsters); R is the outer radius of the rod (mm); N_{rc} is the measured retardation (fringes)

When the above units are used, the unit for axial stress is N/mm². Using the Equations (5.1) to (5.4), the other stresses can be obtained. Figure 5.3 provides a graph which enables the axial stress to be determined for glasses having different stress optical coefficients, given the radius of the rod and the measured retardation at the centre.

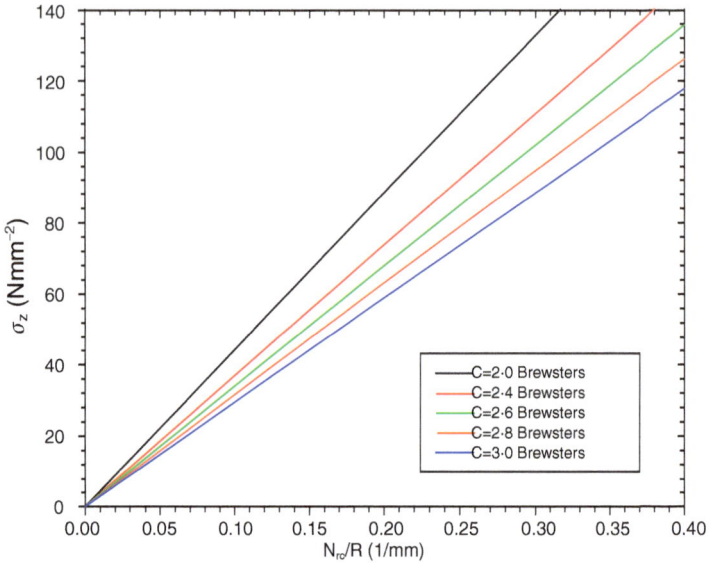

Figure 5.3. Relationship between the axial stress in a rod and the ratio N_{ro}/R for various stress optical coefficients

5.2.2 Tubes

In the case of rods a simple procedure was recommended to obtain a complete picture of the residual stress distribution. Obviously tubes are more complex from this point of view, as additional variables are involved and the temperature histories at the inner and outer surfaces during production (compared with just the outer surface for rods) influence the stress patterns. However, it is possible to obtain the values of stresses at both surfaces of a tube by using semi-empirical relationships established after considerable combined theoretical and experimental studies. Only three measurements are necessary to make use of these relationships and the procedure is as follows.

Initially the correct fluid is selected. For thin-walled tubes proper selection of the immersion fluid is important, because all light rays through the wall thickness are nearly tangential to the outer surface and refraction is therefore very significant. Differences of ±0·005 between the refractive indices of the immersion fluid and the glass produce an error of ±0·02 in the fringe order, which is a typical experimental error for normal photoelastic analysis.

The first stage in measurement is to determine the diameter and wall thickness of the tube to be analysed. It is preferable to make these measurements at the cross-section where the retardation has to be evaluated. With the tube located in the immersion tank as shown in Figure 5.2 then the Sénarmont method can be used to evaluate the maximum negative retardation (N_c) which occurs at the inner surface of the tube. The second retardation measurement is that of the maximum positive retardation N_t which usually occurs at a distance of 0·16t to 0·20t (where t is the wall thickness of the tube) in from the outer surface. A mean value of 0·18t can be used, but if greater accuracy is required then the fractional distance U_m can be determined. The third and final measurement required is that of the location of the point of zero retardation and this is quoted as the ratio of the distance of the point from the outer surface to the wall thickness and is designated by U_o. The stresses can then be obtained from the following formulae:

(a) The maximum compressive stress (on the outer surface)

$$\sigma_c = 0.55 \frac{N_c \lambda}{C\sqrt{RtU_m}} \qquad (5.6)$$

(b) The maximum tensile stress (on the inner surface)

$$\sigma_t = \left(0.92 \left(-\frac{N_t}{N_c} \right)^{1/2} + U_0 - 1 \right) \sigma_c \qquad (5.7)$$

For the tensile stress it should be noted that the maximum value may not always be at the inner surface. In some tubes it occurs

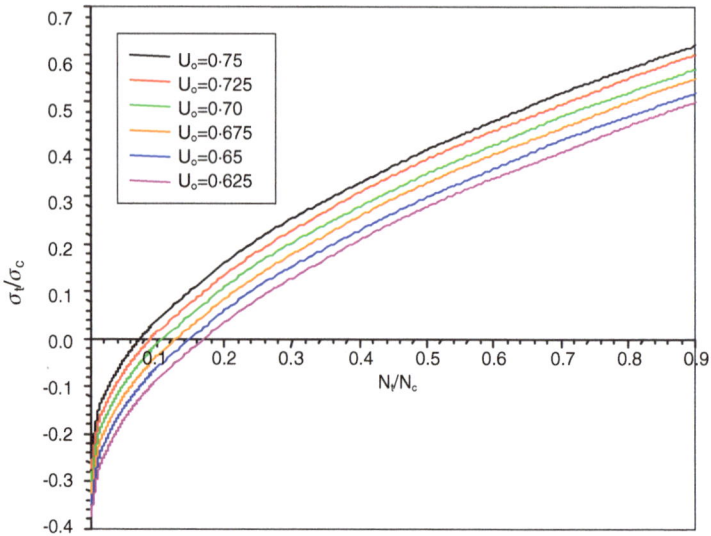

Figure 5.4. Variation of stress ratio with U_o and N_t / N_c

around $1 \cdot 1 U_o$, but the difference between the maximum tension and the surface tension is usually very small.

As an aid, the ratio, σ_t / σ_c, has been plotted in Figure 5.4 for different values of U_o and N_t / N_c.

6. Glass seals

Where one material is sealed to another with a different coefficient of thermal expansion, residual stresses will arise on cooling. Such stresses may compromise the integrity of the seal and it is often useful to be able to assess their magnitude. For either glass-to-glass or glass-to-metal seals the residual stresses in the glass may be evaluated by photoelastic techniques.

Glass-to-metal seals can be described (see Partridge [21]) as being matched, unmatched, soldered or mechanical joints. In matched seals attempts are made to minimise differences in thermal expansion coefficient between the metal and the glass so that residual stresses are kept within a safe limit. In unmatched seals the differences are not minimised but this may not be a problem if the metal component is sufficiently small. Stresses generated in unmatched seals may also be relieved by yielding of the metal component.

High levels of residual stress within a glass seal could lead to failure of the seal and thus normally consideration is given to the measurement of relatively low levels of stress-induced optical retardation such as that shown in Figure 6.1.

To improve the measurement sensitivity, Partridge [21] recommends that a tint plate be used to provide an additional full wavelength retardation in the green region (usually λ=549 nm or 565 nm) to supplement the retardation arising from the residual stresses in the glass. The green region of the spectrum is chosen since the human eye is usually most sensitive to changes in colour in this region. In addition the use of a tint plate means that an immediate judgement can be made as to whether the retardation is positive or negative, as described in Section 3.2.5.

Seals in general have a complex geometry and this leads to correspondingly complex stress distributions within the seal. As seals usually have surfaces which are not optically flat it is necessary to immerse the seal in an appropriate immersion fluid as described in Section 3.4 before photoelastic measure-

Figure 6.1. Retardation in the region of a glass-to-glass seal (full-wave plate in the field)

ments of the seal are undertaken using the methods described in Section 3.2.

In a glass-to-metal seal the principal axis of the metal can be used as a guide to determine the appropriate orientation of the seal in a polariscope for photoelastic measurements. To determine the axial stresses in a rod-type symmetrical seal the polariser should be orientated at right angles to the longitudinal direction of the seal.

A photoelastic analysis for a flat two-wire seal is described by Partridge [21]. The analysis was undertaken on model flat two-wire seals. These seals were assumed to be in a state of plane stress although, in the immediate vicinity of the wires, this is unlikely. The analysis entailed viewing the specimen in a plane polariscope to generate isoclinics and thereby directions of principal stresses. Seals involving several different metal wires were examined and the observed stress distribution was, in general, symmetrical. The values of the principal stresses can also be obtained, as described by Partridge [21], by starting at the boundaries of the seal where it is known that one of the principal stresses must be zero, in a similar manner to that referred to in Section 2.4.2.

7. Special techniques

In addition to the more generally applied techniques described previously, certain special techniques have been developed for particular situations. These techniques may require additional specialist equipment that is not generally available and thus are probably more relevant to laboratory users than those concerned with production control. Two such techniques are described in this chapter.

7.1 Magnetophotoelasticity

As explained in Section 2.4.2, the conventional transmission polariscope is of limited use when a knowledge of the discrete, unaveraged principal stresses is required in, say, a thermally toughened glass where the stresses vary parabolically through the glass thickness. Magnetophotoelasticity (mpe) (Aben [22], [23] and Clarke [24]) offers the possibility of overcoming this difficulty. Mpe is basically a transmission technique of photoelasticity in which the addition of a magnetic field in the region of the glass being studied allows new information to be obtained pertaining to the stress distribution along the direction of view. This additional information is generated as a consequence of the Faraday effect whereby the plane of polarisation of a plane polarised light beam is rotated as it passes through a stress-free glass plate positioned in a magnetic field having its direction parallel to the light passing through the sample. Thus if plane polarised light is passed through stressed glass placed in a magnetic field the resultant rotation is due to both the Faraday and the photoelastic effects. The new data can be used to solve the problem caused by the integrating effects of conventional photoelastic methods, although normal incidence mpe still only yields principal stress differences. However, because the stress differences are no longer averaged, it is possible to apply stress separation techniques, such as oblique incidence, to obtain individual stresses. An mpe polariscope is illustrated in Figure 7.1.

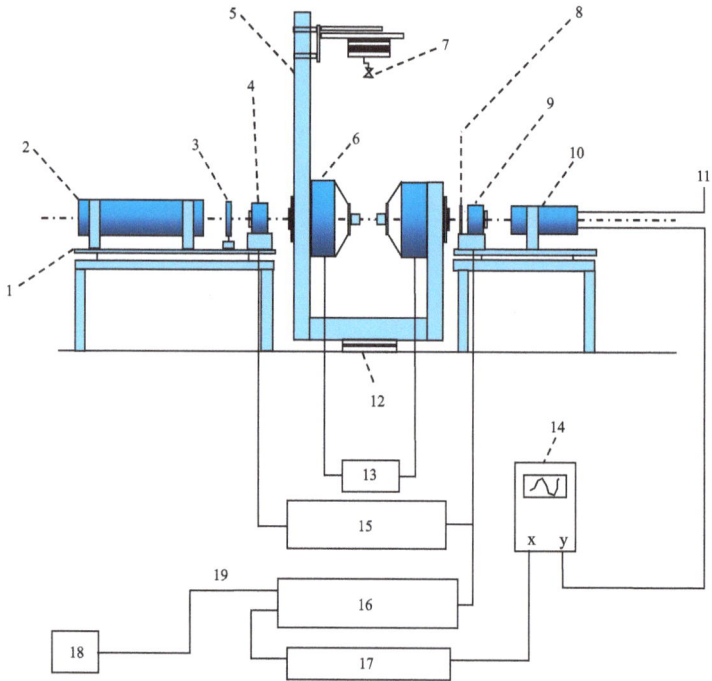

1 Adjustable bench : 2 Helium-neon laser : 3 Lens : 4 Prism polariser : 5 Support and positioning system : 6 Open yoke 178mm electromagnet : 7 Tong for windscreen support : 8 Quarter-wave plate : 9 Prism polariser : 10 Red-extended photomultiplier : 11 Photomultlier power supply : 12 Rotary mount : 13 Magnet power supply : 14 Storage oscilloscope : 15 Electronic coupling : 16 Electronic control unit for polariser and analyser stepping motors : 17 Digital to analogue converter : 18 Data printer : 19 Binary coded decimal output

Figure 7.1. Diagram of magnetophotoelastic polariscope

The theoretical equations for mpe, which were developed by Aben [22] and [23] and described in detail by Clarke [24], can be solved directly only for the constant stress case. However, the solution can be expressed in terms of four parameters for any particular stress distribution in a magnetic field. These parameters are α_1 and α_2, the primary and secondary characteristic directions relative to a principal stress direction, respectively, and α, the angle between α_1 and α_2 ($\alpha = \alpha_2 - \alpha_1$) as shown in Figure 7.2. The fourth parameter Δ^* is the characteristic phase retardation and is the optical retardation in the sample with the magnetic

BASIC OPTICAL STRESS MEASUREMENT IN GLASS

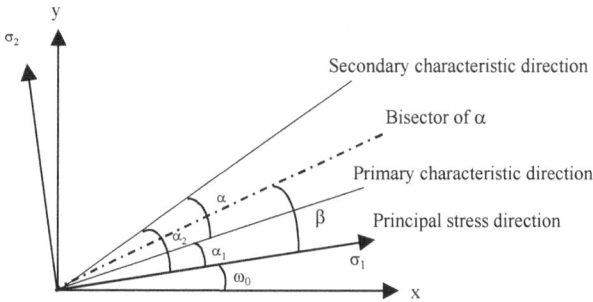

Figure 7.2. Definition of characteristic directions

field applied, being directly analogous to the conventional retardation, Δ, as measured by the Sénarmont technique.

If plane polarised light travelling in the direction of the magnetic field is incident normally on a sample such that the angle between the plane of polarisation and the major principal stress direction is α_1, then it will emerge plane polarised at an angle α_2 to the same principal stress direction. If the glass is unstressed, a rotation of the plane polarised light will still take place due to the Faraday effect alone. It is worth noting that all of the above parameters are dependent upon principal stress differences $(\sigma_1 - \sigma_2)$, as in conventional photoelasticity.

As there is no direct solution for the mpe equations, other than for the constant stress case, an algorithmic technique is used. This entails subdividing an assumed stress distribution into a number of equal thickness layers in which the stress difference is taken to be constant in each. In the case of thermally toughened glass the assumed stress distribution need not be symmetrical. An illustration of the subdivision of a linear bending stress form is given in Figure 7.3.

The analysis proceeds by obtaining a solution for each of the layers, using the constant stress approach, and then combining the solutions for adjacent layers to provide the solution for the overall glass thickness. The stress distribution modelled is

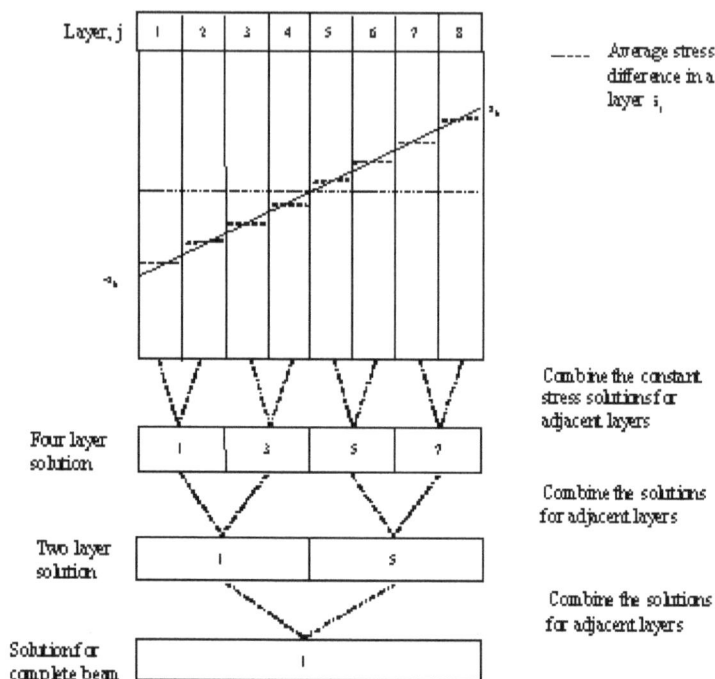

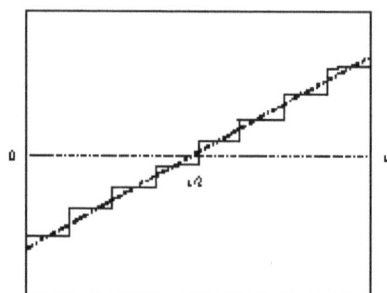

Figure 7.3. Diagrammatic representation of the algorithmic method
of solution for a linear bending stress difference distribution

BASIC OPTICAL STRESS MEASUREMENT IN GLASS

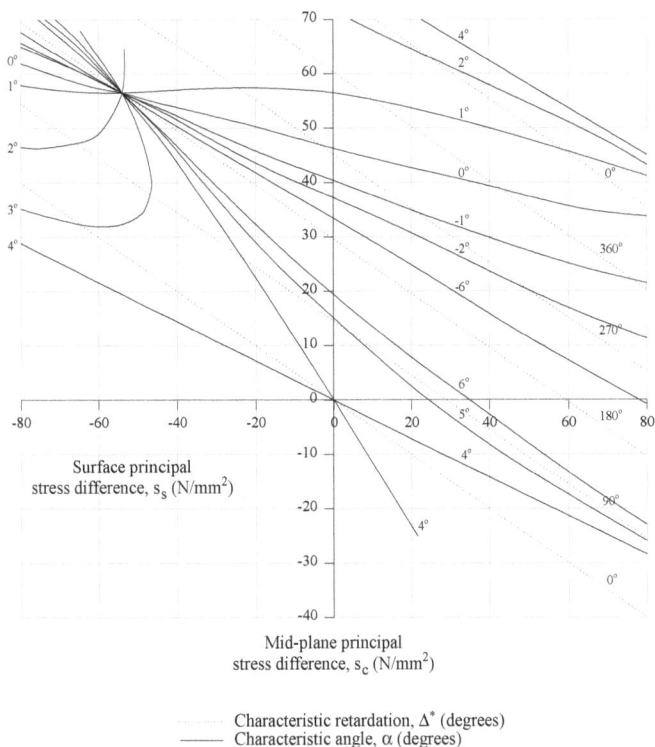

Figure 7.4. Nomogram for toughened glass having a parabolic stress distribution through the thickness; t=6 mm, ψ_t=4·0°

an approximation to the actual continuous system but, by the appropriate choice of number of layers, sufficient accuracy can be achieved. Normally, 16 layers is acceptable. Using such an approach, it is necessary to construct a nomogram which shows the variation of α and Δ^* for various assumed stress distributions and a given magnetic field strength, ψ_t, as illustrated in Figure 7.4.

Values for α and Δ^* can be measured for the same magnetic field strength and then, by referring to the nomogram, the stress differences corresponding to the measured values determined.

In the main, this manual approach is not adopted since it is suitable only for simplest of cases. Rather the analysis is carried out using a PC.

As with all photoelastic techniques such an analysis yields only principal stress differences. In order to separate the principal stress throughout the glass thickness, oblique incidence can be used in conjunction with the above procedure. A combination of normal and oblique incidence stress differences enables the separate stresses to be determined throughout the glass thickness.

7.2 Spectral contents analysis

Most of the methods for measurement of retardation described previously have been manual or semi-manual. However, there are occasions when many measurements have to be made or the level of retardation is low. In both these instances, for different reasons, the scatter in manual measurements at any one position can be wide. Under these circumstances an automatic or semi-automatic approach is desirable. With the advent of the personal computer, there has been considerable effort expended in making the measurement of isoclinics and isochromatics automatic, with varying degrees of success (Redner [25] and Haake & Patterson [26]). The spectral contents analysis technique is one such method of analysis and in its original form was a point-by-point method but, by using a CCD camera a full-field system can be produced.

The method of spectral contents analysis essentially uses the colour of the light provided by the retardation in the glass when it is viewed in a polariscope to make a measure of the magnitude of the retardation. This is done by measuring the intensity in an unstressed glass for different wavelengths of light, provided by narrow band pass filters and a light source having a uniform spectral response over the visible range.

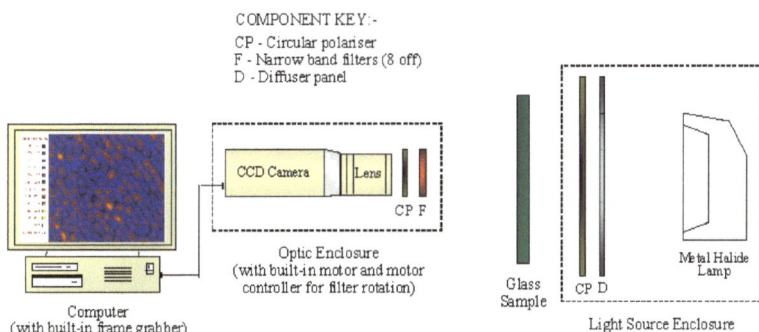

COMPONENT KEY:-
CP - Circular polariser
F - Narrow band filters (8 off)
D - Diffuser panel

CCD Camera | Lens
CP F

Optic Enclosure
(with built-in motor and motor
controller for filter rotation)

Computer
(with built-in frame grabber)

Glass
Sample

CP D

Metal Halide
Lamp

Light Source Enclosure

(a) Diagrammatic representation of layout of apparatus for spectral contents analysis

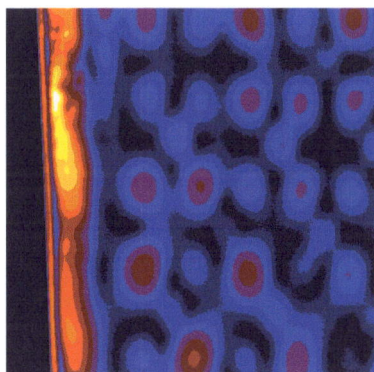

(b) Retardation pattern obtained in a localised region of a toughened glass from a full-field spectral contents analysis

Figure 7.5. Illustration of the apparatus required for spectral contents analysis and the retardation pattern from a localised region of a toughened glass

This is then followed by making similar measurements on the stressed glass at all pixel locations using the same light source and filters. By normalising the intensities at the pixel positions obtained for stressed and unstressed glasses, the transmission at each measurement position is found. Summing the transmissions over the wavelength range at each pixel position enables the overall experimental transmission to be determined. This is then compared with a theoretical transmission summed over the same wavelength range. However, the theoretical transmission includes a retardation term, and to determine the retardation at any pixel position the retardation term is varied until the difference between the summed theoretical and experimental values is a minimum at the chosen pixel location. The value of the retardation when this occurs is the value of retardation at that pixel position. This procedure is repeated at all pixel positions. Due to the considerable amount of data which must be processed to achieve a solution at all points of, say, a 512×512 array (262144 points) it is essential that the operations are carried out on a computer. The apparatus used to make measurements by means of spectral contents analysis is shown diagrammatically in Figure 7.5 (a) while Figure 7.5 (b) gives an illustration of the retardation pattern obtained when spectral contents analysis was used to examine a localised region of a toughened glass.

Acknowledgement

The authors acknowledge the support provided by Pilkington plc and its personnel, in particular, Mrs D. Birch and Dr J. M. Williams, in the preparation of this monograph.

Discussion with personnel from The Ravenhead Company, St Helens and United Glass Limited, St Helens are also much appreciated.

References

1. Frocht, M. M., "Photoelasticity", Vols. 1 and 2, John Wiley and Sons, 1948.
2. Coker, E. M. & Filon, L. N. G., "A Treatise on Photoelasticity", Cambridge University Press, 1931.
3. Kuske, A. & Robertson, G., "Photoelastic Stress Analysis", John Wiley and Sons, 1977.
4. Aben, H. K. & Guillemet, C., "Photoelasticity of Glass", Springer-Verlag, 1993.
5. Hondros, G., "The Evaluation of Poisson's Ratio and the Modulus of Materials of Low Tensile Resistance by the Brazilian (Indirect Tensile) Test with particular Reference to Concrete", *Aust. J. App. Sci.*, (10) 243–268, 1959.
6. Hallimond, A. F., "The Polarising Microscope", Cooke, Troughton and Simms Ltd., York, England, 1953.
7. Smolik, O. & Bellow, D. G., "On the Mixing of Photoelastic Immersion Fluids", *Exp. Mech.*, (14) 400–402, 1974.
8. Singh, S., "An Exact Technique for Mixing Immersion Fluids", *Exp. Tech.*, (7) 27–29, 1983.
9. Collett, E., "Polarised Light - Fundamentals and Applications", Marcel Dekker Inc., New York, 1993.
10. Holister, G. S., "Experimental Stress Analysis", Cambridge University Press, 1967.
11. Dally, J. W. & Riley, W. F., "Experimental Stress Analysis", McGraw-Hill Book Co., 1965.
12. Ansevin, R. W., "The Non-Destructive Measurement of Surface Stresses in Glass", *ISA Transactions*, **4** (339–343) October 1965.
13. Guillemet, C. & Acloque, P., "New Optical Methods for the Determinaton of the Stress Near the Surfaces", *Rev. Franç. Mech.*, No. 4, 157–163, 1962.
14. Kishii, T., "Thermally Tempered Glass Surface Stress Measurement by Critical Ray", *Optics Laser Technol.*, (12) 99–102, 1980.

15. Guillemet, C. & Acloque, P., "Etude des Couches Super-ficielles par Ondes Lumineuses Normales à la Surface". *C.r. Colloque sur la Nature des Surfaces Vitreuses Polies*, Paris 1959, Charleroii USCV 1960, 121–134.

16. Redner, A. S., "A Comparison of Stress Measurement Techniques", *Glass*, July 1993, 268–9.

17. Stress Measuring Device (SMD) Technical Manual, Gaertner Scientific, Revised January 1985.

18. Sutton, P. M., "Stress Measurement in Circular Cylinders", *J. Am. Ceram. Soc.*, **41** (3) 103–109, 1958.

19. Ritland, H. M., "Stress Measurement in Cylindrical Vessels", *J. Am. Ceram. Soc.*, **40** (5) 153–158, 1957.

20. Read, W. T., "An Optical Method of Measuring the Stresses in Glass Bulbs", *J. App. Phys.*, **21** (3) 250–257, 1950.

21. Partridge, J. H., "Glass-to-Metal Seals", Society of Glass Technology, Sheffield, 1949.

22. Aben, H. K., "Principles of Magnetophotoelasticity", Conference on Experimental Stress Analysis and its Influence and Design, Institute of Mechanical Engineers, London, 175–182, 1970.

23. Aben, H. K., "Magnetophotoelasticity - Photoelasticity in a Magnetic Field", *Exp. Mech.*, **10** (3) 97–105, 1970.

24. Clarke, G. P., "Measurement of Residual Stress in Glass using Magnetophotoelasticity", PhD Thesis, University of Nottingham, 1979.

25. Redner, A. S., "Photoelastic Measurements by Means of Computer-Assisted Spectral Contents Analysis", Proc. 5th Int. Conf. on Exp. Mech., Montreal, 421–427, 1984.

26. Haake, S. J. & Patterson, E. A., "Photoelastic Analysis Using a Full-Field Spectral Contents Analyser", Conference on Photoelasticity, New Instrumentation, Materials and Data Processing Techniques, SIRA Communications, London, 1993.

Index